Phase Change Material-Based Heat Sinks

A Multi-Objective Perspective

Phase Change Material-Based Heat Sinks

A Multi-Objective Perspective

Srikanth Rangarajan
C. Balaji

CRC Press
Taylor & Francis Group
Boca Raton London New York

CRC Press is an imprint of the
Taylor & Francis Group, an **informa** business

CRC Press
Taylor & Francis Group
6000 Broken Sound Parkway NW, Suite 300
Boca Raton, FL 33487-2742

© 2020 by Taylor & Francis Group, LLC
CRC Press is an imprint of Taylor & Francis Group, an Informa business

No claim to original U.S. Government works

Printed on acid-free paper

International Standard Book Number-13: 978-0-367-34403-0 (Hardback)

Library of Congress Control Number: 2019951619

Visit the Taylor & Francis Web site at
http://www.taylorandfrancis.com

and the CRC Press Web site at
http://www.crcpress.com

To my parents and my teachers

Srikanth Rangarajan

To all my teachers and students

C. Balaji

Contents

Preface

This book has been written with the objective of making the readers familiar with the exciting developments in phase change material (PCM)-based heat sinks for electronic cooling and is largely based on the author's own, original research. A large amount of work is being currently undertaken worldwide in these areas, with numerous potential applications. The subject is therefore topical and also particularly significant as it leads to the questions of the applicability of this technology for cooling applications.

The above begs the question: if the PCM-based heat sinks themselves are not the ultimate cooling solution, then how do we face thermal challenges in the modern era? Though the jury is still out on this, it is important for the electronics cooling community to first look critically at the advantages offered by the concept of PCM-based heat sinks, to better appreciate the ongoing search for the "optimal heat sinks" and to test some of the available heat sinks for their applicability, before the larger objective of optimal design of heat sinks is accomplished.

This book presents the results of the in-house experimental studies on different types of PCM-based composite heat sinks, followed by determination of optimal configurations that maximize thermal performance. The maximization of thermal performance is quantified in terms of the time to reach a set point temperature for these composite heat sinks, the key quantity of interest in this study. Additionally, the time to solidify back to the original state is also taken as another objective in this study. PCM-based composite heat sinks can be used for the thermal management of portable electronic devices like cellular phones, digital cameras, personal digital assistants, notebooks, and so on, which are not operated continuously over long periods. Such heat sinks are often used in intermittent cycles, where the time to solidify plays a key role in the thermal performance. PCMs can withstand a large number of cycles and are thus ideally suited for repeated use. They are selected based on their heat of fusion and melting temperatures for different applications. The melting temperature of a PCM should be below the maximum operating temperature of the equipment. Most of the PCMs usually have a very low thermal conductivity, a consequence of when heat is dissipated from the electronic equipment into the PCM, even before a significant quantity of the PCM melts, the components may reach unsafe temperatures. By using a high thermal conductivity base material, known as a thermal conductivity enhancer (TCE), in conjunction with a PCM, this challenge can be addressed.

In this book, results of detailed and systematic experimental studies and numerical studies for the case of pin fin, matrix pin fin, and cylindrical heat sinks are reported for candidate PCM-based composite heat sinks subject to different boundary conditions. The phase change material used in the present study is n-eicosane. The TCEs used are aluminum pin fins, matrix-type fins, stems, radial fins, and a heat pipe. In all cases, baseline comparisons are done with a heat sink filled with PCM, but without any fin.

Experiments are conducted to determine the time to reach a set point temperature for aluminum finned heat sinks filled with n-eicosane. Due to the transient nature of the experiments, a matched up numerical model is developed to estimate the heat lost during the actual experiments.

Experimental or numerical results thus obtained were integrated with a feed forward back propagation artificial neural network (ANN) to predict operating times. The ANN prediction was then used as a fitness function in multi-objective algorithms to determine the optimum configuration that maximizes thermal performance (melting and solidification). Three types of TCEs were experimentally investigated: (i) a 72 pin fin with discrete heating, (ii) a matrix fin heat sink, and (iii) a cyclindrical heat sink with stems and radial fins. In all the investigations, the volumes of PCM and TCEs were maintained constant.

First, experimental investigations of heat transfer followed by multi-objective optimization of a PCM-based composite 72 pin fin heat sink subjected to individual heat loading of 4 discrete heaters of equal area were done. The total power of the heaters was fixed, while the individual power was varied within limits. In-house experiments were conducted for different combinations of power. The hotspots imposed by a spatially non-uniform heat flux were seen to have a considerable effect on the melting and solidification cycle and hence on the thermal performance of the heat sink. The diversity in the performance and conflicting nature of both the objectives of the heat sink motivates one to perform a multi-objective optimization using multiple distinct multi-objective algorithms to determine the optimum combination of the discrete power levels, which stretch the charging period and minimize the discharging period of the heat sink simultaneously. The solutions thus obtained were finally validated by conducting an in-house experiment for the optimized configurations. This study established that discrete heating scenarios have to be considered in future optimization of the heat sinks. Additionally, it was found that for problems of this class, a non-dominated sorting genetic algorithm (NSGA-II) show superior performance and hence was used in subsequent studies.

A numerical optimization and experimental investigation of PCM-based composite pin fin matrix heat sink, an alternative to the 72 pin fin heat sink, was carried out. The main objective of this study was to determine the optimized configuration of the matrix type heat sink that would stretch the operation time during the heating cycle and minimize the time during the discharging cycle. The heat sink was made of aluminum. A constant heat flux of 1.9 kW/m^2 was applied at the bottom of the heat sink. The numerical results were matched up with the experimental results to determine the overall heat transfer coefficient with the help of commercially available ANSYS Fluent 14.0 software. For constant power level and constant volume of the PCM, 40 different geometrical configurations of heat sinks were considered and the temperature time histories were obtained for both the charging and discharging cycles by using full three-dimensional simulations of flow and conjugate heat transfer including phase change using Fluent 14.0. The output of these simulations was given as an input to a neural network and multi-objective optimization using NSGA-II and was carried out to determine the optimum configuration of the heat sink, which maximizes the charging period and minimizes the discharging period simultaneously.

Finally, experimental investigation of the effect of fins, gravity, rotational convection, and mass of the PCM on the thermal performance of a heat sink subjected to constant heat flux of $5\mathrm{kW/m^2}$, which translates to 6W at the base, was carried out. Three heat sink configurations with two media, namely air and n-eicosane, were investigated to better understand the role of TCEs on the melting and solidification heat transfer. The heat sink and fins were made of aluminum. Temperature measurements were carried out using calibrated wireless temperature transmission. A LiPo battery, along with a potentiometer circuit, was used to regulate the power input to the heater. The heat sink was subjected to 4 fill ratios (0, 0.33, 0.66, 0.99) of PCM/air, 9 orientations ($0°$, $45°$, $90°$, $135°$, $180°$, $225°$, $270°$, $315°$, $360°$), and 3 rotational speeds (0, 60, 120 rpm) simultaneously. The results confirmed that the thermal performance of the heat sink is a strong function of orientation and rotation at lower fill ratios. The thermal performance was found to monotonically increase with the fill ratio for a finned heat sink, and the unfinned heat sink performed well at lower fill ratios. A critical rotational speed above which the Nusselt number at the heated base is enhanced was found to exist.

Finally, thermosyphons were also numerically examined to understand their potential role as thermal conductivity enhancers.

Symbols

amb	Ambient
Al	Aluminum
ANN	Artificial neural network
BF	Brute force
DC	Direct current
ER	Enhancement ratio
exp	Experimental
GA	Genetic algorithm
GP	Goal programming
IIT	Indian Institute of Technology
MRE	Mean relative error
MSE	Mean squared error
NSGA	Non-dominated sorting genetic algorithm
PCM	Phase change material
PSO	Particle swarm optimization
rpm	Rotations per minute
$\mathbf{R}^2$	Sum of residuals
TCE	Thermal conductivity enhancer

Notation

A_{mushy}	Mushy zone constant
c_p	Specific heat at constant pressure, kJ/kgK
Fo	Fourier number
g	Acceleration due to gravity, $9.81m/s^2$
H	Height of the heat sink, m
k	Thermal conductivity, W/mK
L	Latent heat of PCM, kJ/kg
m	Mass of PCM, kg
p_c	Probablity of crossover
p_m	Probablity of mutation
Q	Heat input, W
q_n	Non-dimensionalized heat input, Q/Q_{max}, W
Q_n	Heat input, W
Q_L	Latent heat stored, kJ
Q_S	Sensible heat stored, kJ
F	Objective function
Ra	Rayleigh number
Ste^*	Modified Stefan number
t_{set}	Time to reach set point temperature, s
t_{ANN}	Time to reach set point temperature predicted by the artificial neural network, s
t_{expt}	Device operating time from the experiments, s
T_{base}	Average heat sink base temperature,$^\circ$ C
T_c	Cooling time, s
T_{ref}	Reference temperature, $^\circ$C
T_{set}	Set point temperature, $^\circ$C
T_∞	Ambient temperature, $^\circ$C
v	Volume fraction
x_i	Input vector
w_i	Weights

Greek symbols

α	Thermal diffusivity, m^2/s
β	Thermal expansion coefficient of PCM, 1/k
η	Distribution index
ν	Kinematic viscosity, m^2/s

| ρ_s | Density of aluminum, kg/m^3 |
| τ | Characteristic time for phase change, Fo. Ste |

Subscripts

∞	ambient
a	air
avg	volume averaged
Al	aluminum
c	charging
d	discharging
eff	effective
f	fluid
h	hot
i	inlet
loss	heat loss through the insulation
o	outlet
s	solid

Acknowledgments

I am forever grateful to IIT Madras for helping me accomplish my research and specifically for allowing publication of my research as a research reference textbook. I am also grateful to my mother Padmini Rangarajan and my father Rangarajan for their patience, unconditional love, and support. Thanks to my wife Rengapriya Srikanth for her patience and help during the preparation of this book. Finally, thanks is also due to the cities of Chennai and Binghamton for providing me peaceful and joyful environments for conducting research and writing this book.

Srikanth Rangarajan

I am forever grateful to IIT Madras for what I am today and specifically for allowing publication of this research as a research reference textbook.

C. Balaji

Authors

Srikanth Rangarajan is currently a post-doctoral researcher at the State University of New York, Binghamton. He got his BE in mechanical engineering (2007–2011) from Sri Sairam Engineering College, affiliated under Anna University. He began his career as a graduate engineering trainee at Wabco India (2011–2012). He joined the master's program (MS) in thermal engineering at IIT Madras (2012). He then upgraded to doctoral studies at IIT Madras in 2013, in heat transfer, specifically in the thermal management of phase change material-based heat sinks, under the guidance of Prof. C. Balaji. He has 7 papers published in international journals. He has also presented 6 papers at international conferences. He has 1 patent filed from his doctoral research work.

C. Balaji is currently a professor in the Department of Mechanical Engineering at the Indian Institute of Technology (IIT) Madras, Chennai, India. At present, Prof. Balaji is the editor-in-chief of the *International Journal of Thermal Sciences*. He graduated in mechanical engineering from Guindy Engineering College, Chennai, India in 1990 (University Gold Medal) and obtained both his MTech (1992) and PhD (1995) from IIT Madras in the areas of heat transfer and thermal sciences. He has been consistently well-rated by students at IIT Madras. He has over 180 international journal publications to his credit and has undertaken several sponsored research projects for the government and industry. He has supervised 29 PhD dissertations and approximately 50 master's theses thus far. He is a recipient of the Humboldt Fellowship of Germany (2005), Young Faculty Recognition Award of IIT Madras (2007), Prof. K.N. Seetharamu Award and Medal for Excellence in Heat Transfer Research (2008), Swarnajayanthi Fellowship Award of the Government of India (2008), Tamil Nadu Scientist Award (2010), and the Marti Annapurna Gurunath Award for Excellence in Teaching at IIT Madras (2013). He is an elected fellow of the Indian National Academy of Engineering. He has authored seven books.

1

INTRODUCTION

1.1 Background

Electronic devices are indispensable in modern life. The popularity of compact electronic gadgets in the consumer market has continued to spiral in the last few decades. As technology in devices keeps exploding day by day, the usage of such devices is increasing at a mind boggling pace. In his fascinating work, "Visions," famous futurist and author Michio Kaku predicts a future where human life is predominantly driven by tiny intelligent gadgets. According to him tiny can be as small as a human nail. On the other hand, over several decades ago Moore argued that the density of electronic components will double itself every 18 months. This statement makes researchers wonder what kind of cooling technologies can be used to keep electronic devices under safe operating temperatures.

From the way the electronic industry has progressed, it is quite evident that Moore and Kaku were somewhat clairvoyant about future challenges in electronic cooling. Nowadays manufacturers design and produce electronic components that have high packaging densities. This is a direct consequence of the ever increasing demand for more and more compact electronics. While modern manufacturing industries are capable of engineering these, their thermal management can often become very challenging. Electronic components, like the human body, require a working temperature for safe operation. The reliability of the electronic devices is highly dependent on their temperatures. Even though CMOS technology seems to replace old bipolar technology, the problem of heating never seemed to decrease. Hence, temperature control is critical to ensure reliability and life of the electronic devices. An interesting way to look at electronic cooling is using a thermodynamic viewpoint. Figure 1.1 shows the thermodynamic anlaysis for a heat sink (closed sytem at constant volume). From the first law of thermodynamics for a pure substance

(i)   The charging cycle,

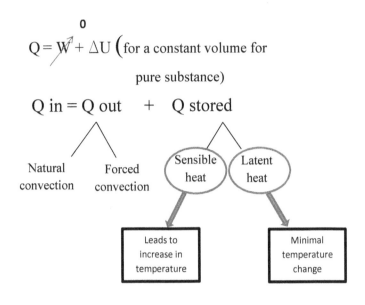

(ii)   The discharging cycle,

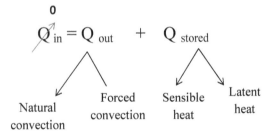

FIGURE 1.1
Thermodynamic viewpoint of electronic cooling.

(say for example phase change material (PCM) placed in a heat sink), it is clear that to maintain a system at an equilibrium temperature the heat input into the system has to be equal to the heat leaving the system under steady state.

If external devices are used to accomplish the cooling, we branch into what is called "active cooling". This book predominantly discusses "passive cooling" which will be defined later in this book.

Even so, the operation of most electronic devices most of the time is highly transient. For this case as seen in Figure 1.1 recourse can be taken to use natural or forced convection for dissipating a part of the heat input. A major brunt of the heat input is then to be borne by the heat storage which can be either sensible or latent heat. The advantages of the latter stem from the high latent heat of melting of PCMs, which helps one to arrest the temperature rise in the device. The cooling load has to balance the heat generated within the component to maintain the whole device at a constant temperature. This idea has driven researchers worldwide to propose different techniques that balance the heat generation and stretch the duration of operation of the devices. Over the last few decades, many researchers have come up with their own solutions to maintain device temperatures within their safe operational limits. Research in the field of cooling of portable electronic devices started with the study of conventional cooling techniques such as air cooling and water cooling. Initial cooling techniques comprised one or more of the following 1. Air natural cooling (convection+radiation) 2. Direct air cooling (Forced convection) 3. Immersion cooling (natural convection) 4. Water cooling (Forced convection) These were considered to be the most effective cooling until the advent of passive type cooling techniques were proposed. A heat flux of 0.5W/cm^2 in an electronic component will result in a temperature difference of 500° C between the device and the ambient air using air cooling, while the prescribed temperature cannot exceed 80° C. In such situations air cooling fails. Forced convection cooling with a constraint on the pressure drop also has not been found to be effective in these situations. This has motivated heat transfer researchers to look for a new method of cooling electronics using a heat sink.

A possible list of questions that may arise in the mind of a researcher is below

1. Why use heat sinks?

2. Why use PCMs within heat sinks?

3. What are the key performance metrics of PCM-based heat sinks?

4. What are the potential design constraints in a heat sink?

5. What is heat sink optimization?

6. Why and what do we optimize?

7. How do we optimize in a multi objective framework?

8. How do we validate the results of the optimization?

This book attempts to answer these questions from the perspective of PCM-based heat sinks.

Heat sinks were termed as a passive type heat exchanger. The advent of heat sinks can be dated back to the early 1930s. Heat sinks facilitate the heat generating elements to dissipate heat into them by providing more heat transfer surface.

Later heat sinks with fins captivated the field with the possibility of dissipating more heat into the surroundings. Even though these heat sinks were cooling the devices effectively, they were based on sensible heating, as a consequence of which the device temperature keeps on increasing with time. A major breakthrough in electronic cooling took place when phase change materials (PCM)-based heat sinks that exploit high latent heat entered the field. These heat sinks were passive in nature.

The operation of PCM-based heat sinks consists of two cycles, namely charging and discharging cycles. The charging cycle or the melting cycle is the process when the heater is switched on and the material undergoes phase change from solid to liquid. The discharging cycle or the solidification cycle is the period when the heater is switched off and the material undergoes phase change from liquid to solid. This book presents the results of PCM-based heat sinks with base heating. The heaters mimic the heat generating electronic components. Now a step by step example from the author's own in house experiment can help the readers better understand the problem statement more clearly. Figure 1.2 shows typical results from a heat sink with air as the internal fluid. There is no flow of air; rather the air is heated in an enclosure. The convection effects are negligible owing to low temperature difference between the wall and fluid.

As expected, the heating A-B and cooling B-D are associated with steep temperature rise (sensible heat storage). Taking a cue from thermodynamic knowledge of latent heat, if air is replaced with a phase change material and the same heating and cooling experiments are carried out the transient temperature history will look like what is shown in Figure 1.3. The result is no different from the heat sink with air case, thus negating the claims that high latent heat alone can deliver thermal management solutions. The heating and cooling cycles exhibit a good amount of sensible heat storage and attendant temperature rise.

This surprising and all the same disappointing result spurred on more intense research to solve the problem. The case with PCM actually becomes a direct victim of the well known "self-insulation" effect. This arises from the low conductivity of the PCM. Fourier's law of heat conduction then came as a godsend to overcome this problem. The idea of using thermal conductivity enhancers(TCE) in conjuction with PCMs in heat sinks was proposed. This is now turning out ot be the state of the art for the heat fluxes in the range of 1 to 50 kW/m^2. The foregoing discussion clearly points to the inadequacy of conventional cooling techniques to meet the demanding cooling requirements during the initial transients that stem from both the compactness and the increasing heat fluxes of such electronic devices. The key goal in the thermal design of any electronic equipment is to keep the device temperature within

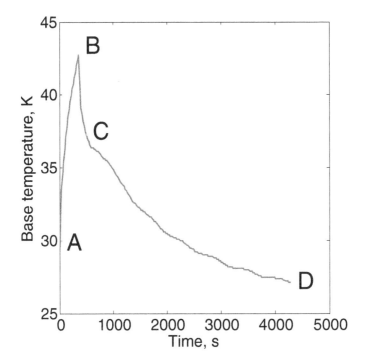

FIGURE 1.2
Typical temperature time history of sensible heating and cooling for an air based heat sink.

safe operational limits to ensure reliability. Phase change material (PCM)-based cooling technique is a passive type cooling that offers several advantages over conventional cooling techniques particularly in intermittent duty cycles due to the large latent heat of the PCM. If a TCE is added to the PCM, as forementioned, the outcome is pleasantly surprising. Pin fins are employed as TCEs. The readers are notified that the role of fins is only to reduce the conduction resistance. The desirable properties of such a PCM are

1. high latent heat of fusion

2. high specific heat capacity

3. high thermal conductivity

4. convenient melting temperature (less than the set point temperature)

5. small volume change during phase change

6. chemical stability

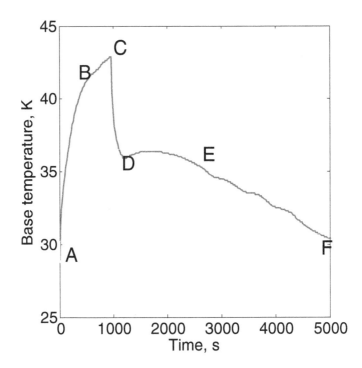

FIGURE 1.3
Typical temperature time history of charging and discharging for a PCM-based heat sink.

Even though high volume expansion can promote solid sinking, to avoid a containment problem during storage inside the heat sink, low volume expansion is more desirable. Furthermore, high volume expansion can induce thermal stresses on the PCM vessel. For example, ice as a latent heat storage material has a very high enthalpy of 334 kJ/kg, but due to its high volume expansion, is not suitable for storage in a cavity like heat sink.

In PCMs, whenever the material changes either from solid to liquid or liquid to gas phase, there is a transfer of energy as material changes its phase from either solid to liquid or liquid to solid. A PCM-based heat sink in the initial phase of operation behaves like a conventional heat sink, and the temperature of the device increases as it absorbs heat. After the PCM reaches its phase change temperature, it has the ability to store latent heat. In other words, it takes up the heat with negligible temperature change. PCM-based heat sinks have the ability to store 5-15 times more heat than the sensible heat storage systems. The major design goal in a PCM-based heat sink would be to stretch the latent heat time during charging and to reduce the discharging cycle time. Latent heat absorption with the melting of a particular PCM can be used to delay the temperature rise of a device subjected to high heat flux. In a PCM-based heat sink, the PCM layer located far away from the heater base can utilize the latent heat only after the adjoining PCM has fully melted. The low thermal conductivity of PCM hence delays the full, uniform melting of the PCM, which in turn results in a high temperature gradient within the PCM heat sink. One way of overcoming the high resistance offered by the PCM to the flow of heat is to add thermal conductivity enhancers(TCE). As already mentioned, the addition of such TCEs ensures a lower temperature gradient and effective utilization of latent heat without the generation of hot spots. The effective thermal conductivity of the PCM-based composite heat sinks is also enhanced.

Unlike sensible heat storage, latent heat storage provides high energy storage density. Since personal digital assistants (PDAs) and other electronic devices are being made with increased compactness, PCM-based cooling is ideally suited for such applications. The operating time of such electronic devices should be equal to the latent heat storage time, so long as the safe temperature limit is in the ballpark of the melting temperature of the PCM (from a time perspective). In practice a little lower and a little beyond the melting sensible heat addition takes place. The catch though is the rapid temperature rise and the phase is often short. This ensures that the temperature of the device is kept in safe operating limits. The PCM used varies from application to application. The cardinal properties governing the selection of PCM are the melting point and latent heat of fusion. The major drawback of a PCM is its low thermal conductivity which results in increased time for both the charging and discharging cycle, and in view of this various techniques are being investigated for improving the thermal conductivity of PCM like fins; metal foams used in conjuction with the PCM are currently being investigated.

Many text books or even the specification tags on electronic devices highlight the maximum operating temperature. Thermal management engineers understand this maximum temperature as the maximum junction temperature. How much can the junction temperature exceed the ambient temperature? This is the key question to be answered. This excess temeperature is often coined as the "thermal budget". There is always a limit for this value. A good thermal management technique has to keep this value as small as possible for it to qualify as the best technique. While readers travel through various chapters in this book they will come across a term known as the "set point", which is the temperature limit for the heat sink.

PCMs are being used in many applications like cooling of mobile phones (PDAs) and reduction of temperature in avionics. Several experimental, analytical and numerical investigations are available in the literature. Detailed reviews on various phase change materials employed in thermal energy storage are available in the literature. Recent advancements in the design of vapor chamber have incorporated PCMs for the initial transients.

Furthermore, PCM-based heat sinks are often complemented with forced convection cooling to extend the time of operation. In a recent work, [41] patented by Sandia Laboratory (USA), the usage of fans is being criticized increasingly in real time applications in the area of electronic cooling. This work reports that heat sink "fouling" is the new root cause of failure. Sandia has already started developing air based impeller type rotating heat sinks. This new development reiterates the fact that rotational type heat sinks will gain currency in the near future.

The readers are notified that a resistance model is not possible for this study as the operation of the heat sink considered is transient and not steady state. Additionally, a one-dimensional resistance model will fail when we consider hotspots.

1.1.1   Multi-objective optimization

The term optimization is defined as the process of finding the best possible (maximum or minimum) solution, given a set of constraints. Optimization is driven by the constraints rather the objective function. Optimization can be broadly classified as single and multiple objectives. Various techniques have been developed to solve both. However in most of the real life applications, one encounters multiple objectives with a set of constraints. These objectives need to be optimized simultaneously, and these objectives are often conflicting in nature. Such problems are encountered in life as for example, (High quality, less price), (Less weight, high performance), (Low pressure drop, high heat transfer). A better way to look at this multi-objective problem is to look at a set of Pareto optimal solutions. If the objectives are not conflicting in nature, then even a multi-objective optimization narrows down to a single solution. It is the conflicting nature of objectives that gives rise to trade offs and hence Pareto optimal solutions. Evolutionary algorithms are quite potent in solving both single and multi-objective problems (MOP). A solution space in a multi-objectives problem may have multiple Pareto curves, but there exists only a single Pareto optimal curve. By definition, the solution to a MOP is Pareto optimal, if there exists no other feasible solution which would degrade any one objective while improving the other . The solution set in the case of a Pareto optimal solution, is generally non-dominant.

In the last two decades, evolutionary multi-objective (EMO) algorithms are being increasingly used. The application of such algorithms is easily seen in several branches of engineering, particularly in problems with conflicting multiple objectives. The origin of EMOs which mimic evolution dates back to the early 1950s. The use of population during the iterations particularly makes the EMOs more suitable for multi-objective optimization, as they are capable of finding multiple Pareto optimal solutions in a single run.

In consideration of the above, the focus of this book is on heat transfer and optimization of melting and solidificiation of phase change material based heat sinks. Fin optimization pertinent to the PCM-based heat sink is different from that of active cooled heat sinks. In active cooling, fin optimization considers increasing the heat transfer and reducing pressure drop. In passive cooling such as PCM-based heat sinks, fin optimization considers achieving uniform melting and quick solidification of the PCM by reducing only the conduction resistance.

1.2 Organization of the book

This book contains nine chapters including the present chapter and a brief description of the contents of each chapter is given below.

Chapter 1 provides a general introduction to phase change material based composite heat sinks. An introduction to thermal management in electronics, active and passive cooling technologies, descriptions about phase change materials and the optimization of thermal systems have been outlined. A brief outline of the organization of the book is also given in this chapter.

Chapter 2 gives a detailed review of the literature of the problems under consideration in the present study. The literature survey is reported for each problem and the summary of the literature survey is presented in a tabular form which is then followed by a statement of the objectives and scope of the present work.

Chapter 3 discusses the characterization of the phase change material n-eicosane and the configurations of the thermal conductivity enhancers employed in this study.

Chapter 4 details the experimental setup and methodology. The instrumentation used in the measurement process is also discussed.

Chapter 5 reports the results of an experimental study on heat transfer of a 72 pin fin heat sink and establishes the motivation for multi-objective optimization.

Chapter 6 reports the results of a comparison of candidate multi-objective optimization algorithms for the problem of discrete heating on a 72 pin fin heat sink.

Chapter 7 provides the results of performance studies on the matrix fin heat sink followed by geometric optimization of this geometry.

Chapter 8 gives the results of experimental studies on cylindrical heat sinks for both the non-rotating and the rotating cases.

Chapter 9 gives the results of numerical studies on heat sinks with thermosyphons as TCEs.

Chapter 10 summarizes the main conclusions of the present study. Suggestions for future studies based on the limitations of the present study are also highlighted.

1.3   Closure

This chapter gave an introduction to the problem of thermal management, various active and passive cooling strategies and phase change materials. An introduction to optimization of thermal systems was also presented. Furthermore, this chapter presented an outline of the way the book is organized. The next chapter presents the state-of-the-art and outlines the scope and objectives of the present study.

2

REVIEW OF LITERATURE

2.1 Introduction

This chapter presents a critical review of the literature pertinent to the present study. For ease and clarity of presentation, the literature review is divided into the following sections.

- Experimental investigations on PCM-based composite heat sinks

- Numerical studies on PCM-based finned heat sinks

- Optimization studies on PCM-based finned heat sinks

2.2 Experimental investigations on PCM-based composite heat sinks

Pillai and Brinkworth [1976] considered the use of phase change material in energy storage systems. They also enumerated the properties of the phase change materials and gave a summary of the selection criteria for the choice of a PCM. The authors also emphasized that design optimization is important for a PCM-based heat sink. The authors recommended choosing a PCM only when it has a significant advantage over sensible heat storage. Using the same

PCM for all applications was heavily criticized and it was concluded that PCM offers a convenient means of storing low grade thermal energy within compact systems.

Zhang et al.[1993] investigated the melting of n-octadecane in an enclosure with discrete heating from the side walls. The solid-liquid interface was tracked using a shawdowgraph technique and melting heat transfer data was reported. The work concluded that the natural convection of the melted PCM has a significant effect on the performance of the heat sink over the unmelted portion. The PCM-based heat sink enhanced the cooling performance by a factor of 2 over the air based heat sink. The optimal arrangement of heaters to improve its performance was discussed. Cho et al. [2003]conducted experiments on micro channel heat sinks to study the effect of hot spots induced by spatially non-uniform heat flux. Configuration of heat sinks that can better remove high heat fluxes of the order 100 W/cm^2 were discussed.

Jones et al. [2006] conducted experimental and numerical studies on melting of a PCM(n-eicosane) inside a cylinder. The heating was done from the side walls. The measured temperatures and the melt front locations of the numerical code with constant temperature boundary condition and the results obtained were validated against the experimental results. The work concluded that at a Stefan number of 0.0836, numerical results show a good agreement with the experimental data. At higher Stefan numbers, the results showed divergence.

Wang et al. [2008] simulated numerically the application of a PCM-based heat sink to the application of mobile electronic cooling. The effect of PCM volume fraction, Stefan number, PCM properties on the thermal performance of the heat sink was studied and it was concluded that higher volume fraction at lower power level gave the best thermal performance.

Luo et al. [2008] considered the specific case of the Samsung SPH E2500 and performed a system level analysis. The heat flux level was taken as 1.9kW/m². The work discussed the scope for using the PCM-based heat sink for this particular application. A thermal resistance model for the heat sink with PCM for that particular mobile phone model was developed.

Dubovsky et al. [2009] studied heat transfer into PCM from an aluminum heat sink with internal fins. The PCM used was RT-35 and the heating was done from the bottom. Numerical simulation were carried out to examine the effect of size and number of fins on the melting of PCM. The main conclusion of this study was that the geometry of the fins play a significant role in the thermal performance of the heat sink.

Fok et al. [2010] considered the cooling of portable electronic devices using phase change material based finned heat sinks. The heat sink was made of aluminum. The authors stressed the effect of ambient temperature on the thermal performance of the heat sink during the discharge cycle. Results also indicated that the usage of the unfinned heat sink was good at low heat inputs. The effect of different power levels on both charging and discharging was studied along with the effect of change in orientation. The work concluded that PCM-based heat sinks prove very efficient in the removal of local hot spots induced in the electronic devices.

Martin et al. [2010] studied the suitability of PCM for cold storage. They discussed the concept of a direct contact PCM water storage for comfort cooling. Two criteria, namely cost and performance, were considered in this study. These were found to be conflicting in nature for all configurations. Experimental studies showed some important drawbacks of using phase change material. These were its low thermal conductivity and volumetric expansion. The work emphasized that the solidification cycle is very slow owing to the low conductivity of PCM which is a major drawback of usage in intermittent cycles.

Yin et al. [2010] built an electronic cooling experimental system based on thermal adaptation composite material. The composite material used was paraffin embedded in graphite. Experimental results showed that the overall heat transfer coefficient is 1.25 times higher than the traditional cooling system. This work criticized the low thermal conductivity of paraffins and their inability for cooling in practical applications. Furthermore, they concluded that the dosage of composite material has a positive impact on the cooling of electronic equipment.

Weng et al. [2011a] performed an experimental investigation on the performance of heat pipes in conjunction with PCM for electronic equipment cooling. This study was analysed from the point of view of the heat pipe side and the heat sink side. The adiabatic section of the heat pipe was covered by a storage container with PCM. The PCM chosen for the current study was tricosane. The work proved that the heat pipe module with PCM could reduce the heater temperature by 12.3^o C as compared to the no PCM case. This could pave the way for the development of heat pipe assisted PCM cooling systems.

Sertkaya et al. [2011], experimentally, investigated the effect of orientation angle on the natural convection heat transfer of pin finned heat sink by including radiation heat transfer. They concluded that there is a considerable effect on the heat transfer when the heating is done from base and negligible when heating from the top.

Baby and Balaji [2012] conducted experimental investigations on PCM-based finned heat sinks using n-eicoanse. The effect of different types of fins such as plate fin and pin fin was studied on the thermal performance of the heat sinks. The work concluded that pin fins are better than plate fins for enhancing the melting heat transfer.

Kozak et al. [2013] performed both experimental and numerical investigations of a hybrid PCM-air heat sink. n-eicoasne was used as the PCM and the heat sink was made of aluminum. The performance of the heat sink at room temperature and elevated ambient temperatures was studied. The work went on to quantify the latent heat and sensible heat content for each configuration studied. The work finally concluded that PCM-based heat sinks are justified only when the latent heat content exceeds that of sensible heat and hence an optimization study is important.

Mahmoud et al. [2013], experimentally, investigated the effect of insert configuration and PCM type on the thermal performance of the heat sink. The investigation concluded that the use of honeycomb type inserts showed better performance compared to machined metallic inserts. Six different PCMs were used and from the results it was seen that the PCM with the lowest melting point was more effective in maintaining the device temperature under safe operating limits.

Fan et al. [2013] conducted experimental investigations to study the performance of the PCM-based heat sinks. For the purpose of baseline comparison, an unfinned heat sink was taken and it was observed that use of a PCM could keep the device temperature under safe operating limits. Two noteworthy conclusions was made (i) High melting point temperature PCMs are preferred for extended operation time and (ii)Fins are mandatory regardless of the PCM employed.

Baby and Balaji [2013], experimentally, investigated the effect of different fin configurations and thereby different melt fractions on the melting of PCM in a cuboidal heat sink. The heat sink was made of aluminum and the PCM used was n-eicosane. An optimization study was also carried out to determine the optimal volume of PCM required to enhance the thermal performance of the heat sink. It was finally concluded that 95% fill ratio was optimal. Single objective heat sink optimization was introduced in a recent work by Baby and Balaji [2019] which could be a precursor for the readers to get a basic introduction on PCM-based heat sinks operation.

Yazici et al. [2019] performed an experimental investigation on the effect on inclination angle and number of fins inside a PCM-based heat sink. The work concluded that the inclination angle of the heat sinks plays a major role in heat sink performance. Amritha et al. [2018] and Srikanth et al. [2018] explored the possibility of multiple PCMs (M-PCMs) in a single heat sink and concluded that at higher heat flux and high set point temperatures, M-PCMs are very effective.

2.3 Numerical studies on PCM-based finned heat sinks

If one follows the literature on numerical investigations it is clear that initially research was directed at developing the right numerical technique to simulate the phase change process. Ho and Viskanta [1984] reported basic heat transfer data during melting from an isothermal wall. n-octadecane was used as the PCM and the cavity shape was rectangular. Shadowgraph technique was used to capture the position of the solid liquid interface position and a numerical model was also developed. Good agreement between the experiments and numerical results was found.

Voller and Prakash [1987a,b] developed an enthalpy-porosity technique for the numerical modelling for solidification and melting including the natural convection. The mushy zone was modelled as a porous medium and the porosity was set equal to the liquid fraction of the corresponding computational grid.

Nayak et al. [2006a,b] developed a numerical model for investigating the effectiveness of using thermal conductivity enhancers to improve the overall thermal conductance of the PCM. Three TCE configurations were studied, namely, rod type fin, plate fin and porous matrix. A transient finite volume method was used to discretise the governing equations. Towards the end of this study it was observed that high volume fraction of TCEs also suppressed the role of convection on the melting process.

Zheng and Wirtz [2004]developed a thermal response model for designing a hybrid thermal energy storage. The work also determined the best design meeting the geometric and heat load constraints by integrating an optimization algorithm with the thermal performance parameters of the heat sink. The final optimized configuration was developed as a prototype and the results were benchmarked.

Bae and Hyun [2004] performed a numerical study on laminar natural convection of air in an enclosure with three flush mounted heaters at the base. This work reiterated the importance of looking at transient temperature change due to discrete heaters for better practical design of electronic devices.

Krishnan et al. [2005] proposed a hybrid heat sink concept that combined active and passive cooling techniques.This work investigated the effect of volume, type of PCM and fin parameters on the thermal performance of the hybrid heat sink. Based on the findings, simple guidelines for the design of the heat sink were developed. The guidelines mainly emphasized the aspect ratio of the heat sink.

Husain and Kim [2008] performed numerical optimization on a micro channel heat sink using evolutionary algorithms. Global Pareto optimal solutions were obtained and analyzed with the available design variables and constraints. This study paved the way for numerical driven search methods and the optimal thus obtained was validated by using the numerical model and a good agreement was found.

Faraji and El Qarnia [2010] performed a numerical study of melting with natural convection inside a cavity with three discrete heat sources of uniform heating. Better designs for PCM-based heat sinks were suggested under discrete heat loading. The work highlighted that some portion of the heat was conducted through the walls before entering the PCM. The major conclusion was that heating from the bottom proved better than heating from the top.

Wang and Yang [2011] performed three dimensional transient numerical simulations to investigate the performance of a hybrid PCM-based heat sink. The number of fins and the heat input values were varied for each simulation. Available experimental data was used to validate the numerical model.

Hosseinizadeh et al. [2011] conducted both experimental and numerical studies on the thermal performance of PCM-based heat sinks. The effects of various parameters like number of fins, fin height and thickness were studied and an optimum configuration of the fins was proposed.

Stupar et al. [2012] modeled a heat sink using an electrical analogy with thermal resistors and capacitors. The validity of the model was verified by conducting experiments. PCM heat sinks with metal foams were also modelled. Later, the model developed was validated with experiments and good agreement was found. However, this model did not find much application in the field, due to large errors in predicting the pressure drop.

Jaworski [2012] explored the thermal performance of a heat spreader containing pin fin like pipes filled with PCM which increases the surface area for convection and thermal capacity due to PCM. Numerical simulations were performed to understand the melting of PCM and he proposed this as an efficient method to remove heat from hot spots in a transient manner.

Zhou et al. [2012] reviewed the storage of thermal energy using phase change material for building applications. The main highlights of the paper included the numerical evaluation of thermal energy storage using PCM in buildings. The criteria for selection of PCM and property measurement methods were also discussed.

Tari and Mehrtash [2013] performed a numerical investigation of natural convection heat transfer from inclined plate fin heat sinks. The authors suggested correlations to be used for the natural convection heat transfer. The numerical model was validated using vertical plate fins on a vertical base heat sink case. The study was concluded with a note that fin spacing played a major role in the heat transfer performance. Zaman et al. [2013], numerically studied the effect of two discrete heaters in an enclosure with air flow. The effect of Rayleigh number on the natural convection process was explored. As the Rayleigh number was increased, the local Nusselt number also increased. However, due to the interacting regimes between two heat sources, there was additional convection to the less heated source.

Bairi and de Maria [2013] quantified transient natural convection in a two dimensional cavity with discrete heat sources. The work presented the results in the form of correlation between Nusselt, Rayleigh and Fourier numbers. Levin et al. [2013] developed a methodology to optimally design a latent heat thermal system. The objective was to minimize the height of the system with two constraints, namely maximum temperature within safe operating temperatures and high latent heat absorption.

Das and Giri [2014] performed a second law analysis of a vertical plate-finned heat sink undergoing mixed convective heat transfer. The mass, momentum and energy equations were solved and the effect on fin spacing on the entropy generation and Nusselt number were presented. The study concluded that there exists an optimum spacing ratio for which Nusselt number is always enhanced for all Grashof numbers.

Halelfadl et al. [2014] performed an analytical optimization study for the total thermal resistance and pressure drop of a rectangular microchannel heat sink. The elitist Non-dominated sorting genetic algorithm (NSGA-II) was used to obtain the set of Pareto optimal solutions. The optimization results were verified with the analytical model and the optimal solutions were found to reduce the total thermal resistance.

Pakrouh et al. [2015] carried out a geometric optimization of a PCM-based heat sink using the Taguchi method. The fin number and fin height were considered as the control variables and optimization was performed on four critical temperatures. The study concluded that the optimum configuration was a strong function of the critical temperature.

2.4 Optimization studies on PCM-based finned heat sinks

The heat sink optimization may be broadly classified as shape/geometric optimization and thermal optimization. Figure 2.1 depicts information flow from the point of view of optimization. However, this book demonstrates both these optimization strategies as separate chapters. But no attempt was made to perform them simultaneously. This can be a possible future study.

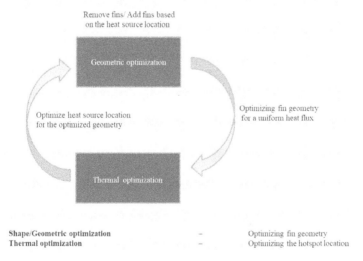

FIGURE 2.1
Flow of information between shape optimization and thermal optimization.

Charnes and Cooper [1977] focused on the implementation of goal and goal interval programming on multi-objective optimization problems. An example problem was also developed and the piece wise linear programming model was applied to this. El-Wahed and Lee [2006] presented an interactive fuzzy goal programming approach to determine the set of non-dominated solutions for a transportation problem. The performance of this approach was evaluated by comparing the results with those obtained with the use of existing algorithms.

Demuth et al. [1992] introduced guidance for the neural network fitting tool with Matlab application. The application was developed with back propagation feed forward network. The tool box thus developed facilitated an increase in the number of neurons in each layer to check the sensitivity on the output of the network thus obtained.

Boggs and Tolle [1995] solved a non-linear optimization problem by implementing the sequential quadratic programming method. The work discussed the practical considerations and applications of this method. Schittkowski [1983] discussed the convergence theory and the implementation of the quadratic programming sub problem. This work also presented a detailed description of a non-linear optimization problem and its solution.

Yoon and Hwang [1995] introduced the Technique for Order of Preference by Similarity to an Ideal Solution (TOPSIS) method which finds its application in multi-criteria decision making problems. The authors described the algorithm with a user-defined weighting vector which can help the decision maker choose one solution among the available set of non-dominated solutions.

Deb et al. [2002] developed an efficient multi-objective algorithm that obtained a more diverse and close to true Pareto optimal solution[1]. The algorithm was tested on standard problems and the results showed better performance compared to existing algorithms.

Akhilesh et al. [2005] proposed a thermal design procedure for composite heat sinks (CHS) for maximising the thermal storage and also to stretch the operation time of the heat sink under safe operational limits. Critical dimensions of the heat sink for which complete melting of the PCM was observed were arrived at.

Husain and Kim [2008] performed numerical optimization on a microchannel heat sink using evolutionary algorithms. Global Pareto optimal solutions were obtained and analyzed with the available design variables and constraints.

Auger et al. [2009] discussed the theory of hypervolume indicator (HI)- a multi-objective optimization performance metric, concluding amongst other equally relevant things that it is not the shape of the Pareto front but rather the slope that determines how the points that maximize the HI are distributed. They also addressed the topic of choice of the reference point so as to obtain the extremes of the front.

Sanaye and Hajabdollahi [2010] performed a multi-objective optimization using non-sorted genetic algorithm (NSGA) to maximize the fin effectiveness

---

[1] A solution which is not dominated by any other solution in the solution space

and minimize the total annual cost incurred to manufacture the fin simultaneously. Six different design variables contributing to the conflict of the two objectives were identified. Additionally, a correlation between the two objective functions and six design variables was developed.

Nadarajah and Tatossian [2010] presented an adjoint method for the multiobjective aerodynamic shape optimization of unsteady viscous flows in a NASA rectangular super critical wing. The optimal that was obtained was validated by conducting experiments. Cuco et al. [2011] employed different multiobjective optimization techniques for a space radiator. Candidate strategies for optimization namely Elitist Non-dominated Sorting Genetic Algorithm (NSGA-II), Multi-Objective Genetic Algorithm (MOGA), Multi-Objective Simulating Annealing (MOSA) and Multi-Objective Generalized Extremal Optimization (M-GEO) were applied to obtain the Pareto front. Furtuna et al. [2011] presented an implementation of the elitist non-dominated sorting genetic algorithm (NSGA-II) applied to the multi-objective optimization of a polysiloxane synthesis process. The work concluded that the algorithm was able to find the entire non-dominated front.

Chaudhry et al. [2011] discussed the optimal positioning of the wireless sensors using a Pareto based evolutionary algorithm called FLEX. The work concluded that the algorithm is more flexible and that the concept of elitism is preserved in every iteration.

Huang et al. [2011] examined a three-dimensional heat sink module design problem to estimate the optimum design variables. The Levenberg-Marquardt Method (LMM) was used along with a general purpose commercial code. The main objective of the study was to determine the best shape of the heat sink that would minimize the maximum temperature of the heat sink.

Ahmed et al. [2013] performed multi-objective optimization and decision making in the selection of a cricket team. The authors tried to meet two objectives, to maximize the batting average and minimize the bowling average, the finite budget was applied as a constraint and a set of non-dominated solutions was obtained.

Jang et al. [2014] conducted multi-objective optimization studies on a heat sink for LED cooling applications. Both convection and radiation effects were considered. They also conducted experiments to validate the numerical model. The two objective functions considered were (i) minimization of the thermal resistance and (ii) minimization of the mass of the heat sink. Finally, a set of non-dominated solutions was obtained that satisfied both the objective functions.

Sun et al. [2014] proposed a technology that combines phase change materials with a cold source for telecommunication base stations. A full prototype was built and tested in the laboratory. They concluded that the proposed technology helps to reduce a significant amount of energy.

Asadi et al. [2014] conducted multi-objective optimization studies for building retrofit. Optimization was performed using a coupled ANN-GA approach and set of non-dominated solutions was obtained.

Dominguez et al. [2014] applied the MOPSO technique for the reduction of energy consumption in electrical lines. The fitness of the algorithm was compared with the true Pareto front and the results of NSGA-II.

Riquelme et al. [2015] reviewed and analysed 54 multi-objective optimization metrics and discussed the benefits and weaknesses and concluded that the hypervolume index was the most commonly used metric. Their analysis was meant to help researchers pick an appropriate performance metric for their work.

2.5   Thermosyphon assisted melting of PCM

Various techniques like fins and heat pipe have been adopted in conjunction with PCMs to enhance the thermal conductivity of the Latent Heat Thermal Energy Storage (LHTES) over the years. A thermosyphon is a gravity assisted wickless heat pipe containing a working fluid, which undergoes phase change during the transient operation of the heat pipe. The evaporator section of the heat pipe filled with water absorbs the major portion(>80%) of the heat input. The condensation at the condenser serves as the heat source for the PCM. Ho and Viskanta [1984] reported the basic heat transfer data during melting from an isothermal vertical wall. n-octadecane was used as a PCM and cavity shape was rectangular. Shadowgraph technique was used to capture the position of the solid liquid interface position and a corresponding numerical model was also developed. Good agreement between the experiments and numerical simulations was found. Webb and Viskanta [1985] attempted to determine the characteristic length scale for correlating melting heat transfer data from a vertical wall. They concluded that height of the enclosure should not be used as the characteristic length scale for melting heat transfer correlations. The authors also suggested the usage of length in the direction of temperature gradient as the characteristic length. Bejan [2013] discussed the basic natural convection correlations from a vertical wall under isothermal boundary conditions. Voller and Prakash [1987a,b] developed a generalized methodology for the modeling of a mushy region phase change problem. They concluded that the Darcy source terms and the latent heat source term are the driving parameters in the fixed grid modeling approach. The main objective of the study was to present an enthalpy formulation based fixed grid methodology for a phase change problem with convection. Farsi et al. [2003], experimentally, studied the transient behavior of a two-phase thermosyphon. The authors also developed an analytical model for the thermosyphon systems and a good agreement between the model and the experiment was seen.

Liu et al. [2006], experimentally, studied the heat transfer characteristics of a heat pipe with a LHTES during both the charging and the discharging cycles. The authors also performed a simplified thermal resistance analysis and matched the numerical results with those of experiments. Shabgard et al. [2010] developed a thermal resistance model to analyze the high temperature LHTES. The authors analyzed the effect of insertion of multiple heat pipes in a PCM enclosure for solar applications and concluded that the addition of heat pipes enhanced the performance of LHTES. Robak et al. [2011], experimentally, investigated the enhancement in the performance of a latent heat storage system due to addition of heat pipes and fins. The performance of the heat pipe was evaluated during both charging and discharging cycles. The authors concluded that the melting rates are enhanced by 60% by the inclusion of heat pipe over fins. Weng et al. [2011], experimentally, investigated the application of heat pipe with PCM for electronic cooling. The evaporator section of the heat pipe takes up the heat generated by the electronics and is discharged through the condenser into the PCM. Three kinds of PCMs were investigated and the authors concluded that the use of tricosane as PCM proved to be more efficient for the heat pipe operation.

Wang et al. [1999], experimentally, investigated the melting process in the vicinity of a heated vertical wall inside a rectangular enclosure. The authors developed a correlation for the melt fraction with Fourier number (Fo), Nusselt number (Nu), Rayleigh number (Ra). Tiari et al. [2015] numerically studied the effect of melting of high temperature PCM assisted by mutiple finned heat pipes inside a square enclosure. The authors concluded that an increase in the number of HPs increased the charging cycle time of the LHTES. Furthermore, they reported that convection plays an important role during the melting process. Srikanth et al. [2015a,b] performed a geometric optimization on a PCM-based composite heat sink with metallic fins as thermal conductivity enhancers (TCEs). The authors concluded that the spacing between the fins (TCEs) in all three directions has a significant effect on both the charging and discharging cycle times. The authors stressed the fact that the spacing that favored the charging cycle did not perform well during the discharging cycle.

From the review of the literature, it is very much evident that the heat pipe-PCM integrated energy storage systems research is a topic of serious research in recent times. Furthermore, numerical simulations rarely capture the transient response of the heat pipe.

For convenience and for easy reference, the salient features of the literature review are presented in the form of Table (see Table 2.1)

2.6   Scope and objectives of the present study

The preceding review shows that even though a considerable amount of literature is available on PCM-based heat sinks, literature on the optimization of the heat sink considering both melting and solidification is scarce and an

TABLE 2.1: Summary of literature review

Sl No	Authors	year	Remarks
(a) Experimental investigations on PCM-based composite heat sinks			
1	Pillai and Brinkworth	1976	Considered the use of phase change material in energy storage systems. They also enumerated the properties of the phase change materials and gave a summary of the selection criteria for the choice of a PCM.
2	Zhang et al.	1993	Compared the effectiveness of passive cooling by PCMs with that of active cooling. Found that the natural convection of the melted PCM has significant effect on the performance of the heat sink over the unmelted portion.
3	Cho et al.	2003	Conducted experiments on micro-channel heat sinks to study the effect of hot spots induced by spatially non-uniform heat flux. Different configuration heat sinks for high heat flux cooling were presented.
4	Jones et al.	2006	Conducted experimental and numerical studies on melting of a PCM(n-eicosane) inside a cylinder. It was concluded that at a Stefan number of 0.0836, numerical results had a good agreement with the experimental data.
5	Wang et al.	2008	The application of a PCM-based heat sink to mobile electronic cooling was studied. The study concluded that higher volume fraction at lower power level gave the best thermal performance.
			Continued on next page

Sl No	Authors	year	Remarks
6	Luo et al.	2008	Thermal resistance model was developed for the heat sink with PCM for a particular mobile phone model.
7	Dubovsky et al.	2009	Studied heat transfer into PCM from an aluminum heat sink with internal fins. The main conclusion of this study was that geometry of fins played a significant role on the thermal performance of the heat sink.
8	Fok et al.	2010	Considered the cooling of portable electronic devices using phase change material based finned heat sinks. The authors concluded that PCM-based heat sinks proved very efficient in the removal of local hot spots induced in the electronic devices.
9	Martin et al.	2010	Studied the suitability of PCM for cold storage. The authors emphasized that the solidification cycle is very slow owing to the low conductivity of PCM which is a major drawback of usage in intermittent cycles.
10	Yin et al.	2010	Built an electronic cooling experimental system based on thermal adaptation composite material. It was concluded that the dosage of composite material has a positive impact on the cooling of electronic equipment.
11	Weng et al.	2011a	Performed an experimental investigation on the performance of heat pipes in conjunction with PCM for electronic equipment cooling. The study paved the way for the development of heat pipe assisted PCM cooling systems.

Continued on next page

Sl No	Authors	year	Remarks
12	Sertkaya et al.	2011	Experimentally investigated the effect of orientation angle on the natural convection heat transfer of pin finned heat sink. It was concluded that there is a considerable effect on the heat transfer when the heating is done from the base and negligible when heating from the top.
13	Baby and Balaji	2012	Investigated PCM-based finned heat sinks using n-eicoanse. The study concluded that pin fins are better than plate fins for enhancing the melting heat transfer.
14	Kozak et al.	2013	Investigated hybrid PCM-air heat sink with n-eicosane as PCM. The study concluded that PCM-based heat sinks are justified only when the latent heat content exceeds that of sensible heat and hence an optimization study is important.
15	Mahmoud et al.	2013	Investigated the effect of insert configuration and PCM type on the thermal performance of the heat sink. From the study it was seen that the PCM with lowest melting point was more effective in maintaining the device temperature under safe operating limits.
16	Fan et al.	2013	Conducted experimental investigations to study the performance of the PCM-based heat sinks. Fin was found to be mandatory regardless of the PCM employed.

Continued on next page

Sl No	Authors	year	Remarks
17	Baby and Balaji	2013	Experimentally investigated the effect of different fin configurations and thereby different melt fractions on the melting of PCM in a cuboidal heat sink. The work finally concluded that 95% fill ratio was optimal.
18	Srikanth et al.	2018	Experimentally investigated the multiple PCMs in a single heat sink and concluded that at high heat flux multiple PCMs are beneficial.
19	Baby and Balaji	2019	Experimentally driven single objective optimization was performed on a PCM-based heat sink.
(b) Numerical studies on PCM-based finned heat sinks			
1	Ho and Viskanta	1984	Reported basic heat transfer data during melting from an isothermal wall.
2	Voller and Prakash	1987a	Developed an enthalpy-porosity technique for the numerical modelling for solidification and melting including natural convection.
3	Nayak et al.	2006a	Developed a numerical model for investigating the effectiveness of using thermal conductivity enhancers to improve the overall thermal conductance of the PCM.
4	Zheng and Wirtz	2004	Developed a thermal response model for designing a hybrid thermal energy storage. The model was found to validate well with the experiments.
5	Bae and Hyun	2004	Found that PCM-based heat sinks are effective for the cooling of mobile phones under intermittent moderate usage conditions.
Continued on next page			

Sl No	Authors	year	Remarks
6	Krishnan et al.	2005	Investigated the effect of volume, type of PCM and fin parameters on the thermal performance of the hybrid heat sink numerically.
7	Nayak et al.	2006a	Performed a numerical study on laminar natural convection of air in an enclosure with three flush mounted heaters at the base. The authors concluded that the natural convection due to discrete heating is significant.
8	Husain and Kim	2008	Performed numerical optimization on a micro-channel heat sink using evolutionary algorithms. Global Pareto optimal solutions were obtained and analyzed with the available design variables and contraints. Optimal thus obtained was validated by using the numerical model and a good agreement was found.
8	Faraji and El Qarnia	2010	Performed a numerical study of melting with natural convection inside a cavity with three discrete heat sources of uniform heating. The major conclusion was that heating from the bottom proved better than heating from the top.
9	Wang and Yang	2011	Performed three-dimensional transient numerical simulations to investigate the performance of a hybrid PCM-based heat sink. The numerical data was validated with experiments and good agreement was found.
10	Jaworski	2012	Performed numerical simulations to understand the melting of PCM assisted with TCEs. Optimal volume of fins for better melting performance was arrived.
			Continued on next page

Sl No	Authors	year	Remarks
11	Zhou et al.	2012	Discussed the criteria for selection of PCM and propert measurement methods.
12	Tari and Mehrtash	2013	Performed a numerical investigation of natural convection heat transfer from inclined plate fin heat sinks. The study concluded with a note that fin spacing played a major role in the heat transfer performance.
13	Zaman et al.	2013	Numerically studied the effect of two discrete heaters in an enclosure with air flow. The authors numerically concluded that as the Rayleigh number was increased, the local Nusselt number also increased.
14	Bairi and de Maria	2013	Found that PCM-based heat sinks are effective for the cooling of mobile phones under intermittent moderate usage conditions.
15	Levin et al.	2013	Proposed an optimization procedure for the design of a latent heat thermal energy storage system. The objective was to minimize the height of the system with two constraints, namely, maximum temperature within safe operating temperatures and high latent heat absorption.
16	Das and Giri	2014	Performed a second law analysis of a vertical plate-finned heat sink undergoing mixed convective heat transfer. The study concluded that there exists a optimum spacing ratio for which Nusselt number is always enhanced for all Grashof numbers.

Continued on next page

Sl No	Authors	year	Remarks
18	Pakrouh et al.	2015	Found that PCM-based heat sinks are effective for the cooling of mobile phones under intermittent moderate usage conditions.
(c) Optimization studies on PCM-based finned heat sinks			
1	Charnes and Cooper	1977	Implemented goal programming to a multi-objective optimization problem.
2	Boggs and Tolle	1995	Implemented sequential quadratic programming to the goal programming application.
3	Schittkowski	1983	Defined the convergence criteria for a multi-objective optimization problem.
4	Yoon and Hwang	1995	Compared candidate multi-objective optimization algorithms to test problems.
5	Deb et al.	2002	Introduced the state of the art NSGA-II algorithm for multi-objective optimization problems.
6	Husain and Kim	2008	Performed numerical optimization on a micro-channel heat sink using evolutionary algorithms.
8	Cuco et al.	2011	Employed different multi-objective optimization techniques for a space radiator problem. NSGA-II was concluded to be more efficient among the candidate multi-objective algorithms.
9	Furtuna et al.	2011	Performed multi-objective optimization of a polysiloxane synthesis process concluded that NSGA-II was able to capture the Pareto front completely.
10	Jang et al.	2014	Conducted multi-objective optimization studies on a heat sink for LED cooling applications. The optimal solutions obtained were validated by conducting experiments.
Continued on next page			

Sl No	Authors	year	Remarks
11	Riquelme et al.	2015	Reviewed and analysed 54 multi-objective optimization metrics and discussed the benefits and weaknesses and concluded that the hypervolume index was the most commonly used metric.
(d) Thermosyphon assisted melting of PCM			
1	Farsi et al.	2003	Experimentally studied the transient behaviour of two phase thermosyphon.
2	Liu et al.	2006	Experimentally studied the performance of heat pipe with LHTES in both the charging and discharging Cycles.
3	Robak et al.	2011	Experimentally studied the performance enhancement in LHTES due to addition of fins/heat pipe
4	Tiari et al.	2015	Numerically investigated the melting of PCMs assisted with thermosyphons

optimization strategy based on experimental data is scarcer. In consideration of the above, the objectives of the present study are as follows.

1. Experimental investigations on heat transfer performance of a 72 pin fin heat sink with discrete heating at the base.

2. Development of a mathematical model to predict the time to reach a set point temperature considering both melting and solidification cycles followed by thermal optimization studies.

3. Critical evaluation of candidate multi-objective optimization algorithms for the problem of discrete heating in a 72 pin fin heat sink.

4. Experimental investigations on heat transfer performance of a matrix heat sink, followed by multi-objective geometric optimization.

5. Experimental investigations on heat transfer performance of a cylindrical heat sink subject to different fill ratios, orientations and rotation.

6. Numerical investigation of thermosyphon as a potential thermal conductivity enhancer in a PCM-based heat sink.

2.7   Closure

A detailed reviewed of the literature pertinent to the problems considered in this study was presented in this chapter. The scope and objectives of the present study were laid out. The next chapter presents the details of the characterization of the phase change material and the thermal conductivity enhancers.

3

CHARACTERIZATION OF PCM AND TCEs

3.1 Introduction

This chapter deals with the characterization of the PCM and the TCE. In a PCM-based heat sink, the thermal control unit is composed of PCM and TCE. The n-eicosane used as the PCM in the present study is supplied by Sigma Aldrich, USA. The cost of 100 g of n-eicosane is 55 USD (2019 prices). In the subsequent sections, an elaborate discussion of the properties required for the selection and the characterization strategies employed for the PCMs and TCEs are presented.

3.2 Selection of phase change material

Phase change thermal control devices have been discussed extensively in the literature. These articles often refer to a device of this type by different names such as thermal capacitor, thermal flywheel, heat of fusion device, latent heat device, and fusible mass device. However, all of these terms refer to a component which is used to either thermally control a medium or store energy by utilizing a material which undergoes a change of phase. There are a number of phase change transformation classes such as

- Solid-liquid transformations

- Liquid-gas transformations

- Solid-gas transformations

- Solid-solid transformations

- Liquid-liquid transformations

A negligible amount of energy is released or absorbed by liquid-liquid transformations, and it is questionable whether this is a true class of phase change. In this work n-eicosane, a solid-liquid phase change material is employed. Solid-liquid transformations are of great importance because most classes of materials undergo this type of transformation without exhibiting large volume changes while releasing or absorbing relatively large quantities of energy.

Since hydrocarbon properties are dependent on purity, differences in reported property values may be directly related to differences in purity, a quantity often not specified. In this study, n-eicosane with 99% purity is used ([35]). It is understood that the term paraffin generally denotes any of the saturated aliphatic hydrocarbons of the methane series C_n H_{2n+2}. In the present study, n=20 (n-eicosane).

Thermal energy can be stored as latent heat in which the energy is stored when a substance changes from one phase to another by melting or freezing. During phase change, the temperature of the substance remains constant. When PCMs are applied in latent heat thermal storage units, they undergo a change of phase, as for example from solid to liquid or vice versa during the energy transfer process. Even though over 500 potential PCM candidates are available mainly under the organic, inorganic and eutectic category, only very few materials have actually been tested and used for PCM applications such as telecom shelters, transportation, solar thermal applications, air conditioning systems, passive heating of buildings, textile industry, electronics, green houses, temperature peak stabilization and so on. The PCMs are selected mainly on the basis of their heats of fusion and melting temperatures. Many PCMs have a high latent heat of fusion and a convenient melting temperature. Even so they are corrosive or hazardous. For use in practical applications, PCM must exhibit certain desirable thermodynamic and chemical properties. A detailed review on various PCMs, their properties and applications is available in ([35]). In this report, various criteria that determine the selection of the PCM are given. Even though a PCM is selected based on applications, it should meet the following requirements.

- The melting temperature of the PCM should be lower than the device's maximum temperature.

- The latent heat of fusion must be high, especially on a volumetric basis to minimize the size of the heat storage unit.

- High thermal conductivity must be available that would make the PCM melting and solidification homogeneous and also provide the capability of preventing the potential PCM overheating.

- A high specific heat must be available to provide additional sensible heat storage capacity.

- A high density must exist to allow a smaller size of storage container.

- To reduce the containment problem, the PCM volume changes on phase transitions and the vapor pressure at operating temperatures should be minimal.

- The PCM must have chemical stability, so that it will be suitable for repeated use. It should be compatible with materials of construction and should be non-toxic.

- No or very small super cooling should exist.

- PCM should be abundant, available and cost effective.

Latent heat thermal energy storage is very attractive mainly due to its ability to provide high energy storage density and its characteristics to store heat at constant temperature corresponding to the phase transition temperature of the PCM. The phase change can be solid-liquid, solid-gas, liquid-gas and vice versa. The thermal properties of n-eicosane are provided in the NASA design handbook [35].

3.2.1 Sensible and latent heat time

For any configuration of heat sink with PCM, the latent heat time of the heat sink is taken to be $36.5°C - 38.5°C$, because exact isothermal phase change does not happen in reality; rather it happens with minimal temperature difference. The temperature range of the phase-change process for n-eicosane is $36 - 38°C[3] - [4]$.

3.3 Thermal conductivity enhancer (TCE)

Because of the high thermal conductivity of metals such as aluminum, copper, titanium and stainless steel, they are preferred to be used as TCE for various applications. Heat sinks can be made from milling, electrical discharge machining, extrusion or casting. Usually heat sinks used in electronics are fabricated from copper or aluminum because of their higher thermal conductivities. Low density and corrosion resistance make aluminum the material of choice for the TCE, especially in the thermal management of portable electronic equipment. A metal joining process like soldering and brazing can be easily done with aluminum. In addition, aluminum-aluminum joints are generally superior in strength to joints between aluminum and dissimilar metals.

TABLE 3.1
Properties of materials employed in the present study

Material	Thermal conductivity (W/mK)	Specific heat (kJ/kgK)	Latent heat (kJ/kg)	Melting point (°C)	Density (kg/m³)
n-eicosane	0.39 (solid) 0.157(liquid)	1.9(solid) 2.2(liquid)	237.4	36.5	810(solid) 770(liquid)
Aluminum	202.4	0.87	-	660.4	2719
Cork	0.045	0.350	-	-	120

Even though the thermal conductivity of copper is double that of aluminum, copper has a much higher density, which is well over three times the density of aluminum. This weight penalty makes copper not quite suitable for the type of applications involving the use of low weight heat sinks. Reiterating the fact that these heat sinks are employed in transient operation, the mass of the heat sink is important. For a steady state operation, the mass of the heat sink is seen to play no significant role.

The properties of the materials employed in this study are shown in Table 3.1.

3.4 Closure

This chapter reported the characterization studies of the TCEs and PCMs. Additionally, plots available from the design handbook by [35] for properties of n-eicosane were presented for the sake of completeness. Details of the experimental setup and methodology and the instrumentation used in the present study are given in the next chapter.

4

EXPERIMENTAL SETUP AND INSTRUMENTATION

4.1 Introduction

The experimental setup and measuring instruments used for the present work, the experimental methodology and the uncertainty in the measurements, are detailed in this chapter. The experimental setup needs to be designed in such a way to reduce the energy loss such as conduction through the base during heat transfer and eliminating disturbances during experiments. Since a class of problems is considered in this study, different setups are required for accomplishing the objectives. Descriptions of these setups are provided in the ensuing sections.

4.1.1 Heat sink design

The heat sink used in the present study is made of aluminum. The PCM considered in this study is n-eicosane. A picture of one of the heat sinks is shown in Figure 4.1. The 72 pin fins are fabricated using the electro discharge machining process. The heat sink cavity has a depth of 20 mm. The fins have a cross sections of 2 mm ×2 mm. The heat sink has a wall thickness of 7mm. All four sides of the walls and the bottom face of the heat sink are covered with cork to minimize heat loss to the ambient. The 72 pin fin heat sink has proven to be the optimal geometry in the work by Baby and Balaji (2019).

A 2 mm slot is given at the bottom face of the heat sink to accommodate the heaters. The top face of the heat sink is covered with acrylic by means of bolts. A silicon gasket of 2 mm thickness is placed between the acrylic and the top face of the heat sink to avoid any spillage or leak of the PCM. A reasonable torque on the bolts is essential to prevent any PCM leak. To mimic discrete heat dissipating units, four discrete heaters are placed at the bottom of the heat sink. The positions of the discrete heaters are shown in Figure 4.2. The

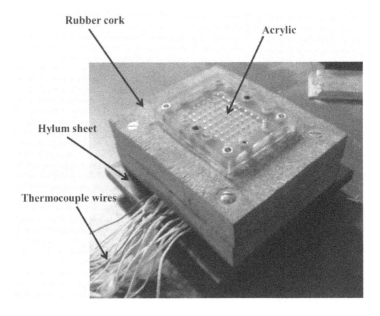

FIGURE 4.1
Photograph of the PCM-based heat sink with 72 pin fins after assembly [65].

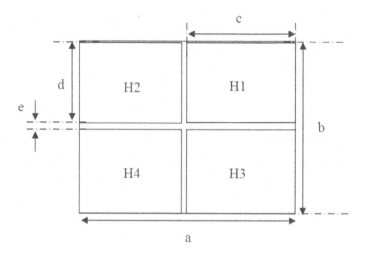

FIGURE 4.2
Schematic of the discrete heaters [63].

heaters are made of mica sheet over which the nichrome wire is wound. The size of the mica sheet is 29 mm × 20 mm.

The heater area available at the base is 60 × 42 mm². This area is subdivided into four equal areas into which the discrete heaters are placed. The dimensions of the heaters are the same as in [64] and are shown in Figure 4.2 and Table 4.1.

A gap of 1 mm is maintained between the heaters to avoid any electrical contact. The diameter of the nichrome wire is 0.2 mm. Nichrome is chosen as it is highly corrosion resistant and does not oxidize at higher temperatures. The nichrome wire is coiled in a zig zag manner over the mica sheet. The output of the heater design process is the length of the heater. The resistivity of the nichrome wire used in the present study is 1.5×10^{-6} Ω- m. The heater is attached to the base of the heat sink using a thermal interface material (TIM) of 5 W/mK thermal conductivity. The heat generated by the heating element depends on the current passing through the wire, which in turn depends on

TABLE 4.1
Heater dimensions for Figure 4.2 [63].

SI.No	Dimension	Value in mm
1	a	60
2	b	42
3	c	29
4	d	20
5	e	2

the resistance, the length, and the cross section of the wire. The DC power source used in the present study has a voltage rating of 0-10V and a current rating of 0-1.25 A. The power P is given by

$$P = \frac{V^2}{R} \qquad (4.1)$$

To obtain a minimum power of 0.5 W and a maximum power of 2 W, the length of the nichrome wire for each discrete heater was calculated to be 440 mm.

4.1.2   Thermocouple positions

For the present study, a total of 16 calibrated K-type thermocouples are used for temperature measurements at various locations in the heat sink. Calibration is carried out using a constant temperature bath in the temperature range of 30 to 70° C. Six thermocouples, as shown in Figure 4.3, are placed at the base of the heat sink.

One thermocouple is placed at the centre of each discrete heater. In the case of a single plate heater, only two thermocouples are placed at the centre of the heat sink as shown in Figure 4.4. Two thermocouples are placed at the centre of the base of the heat sink. Four thermocouples are placed on the walls. To

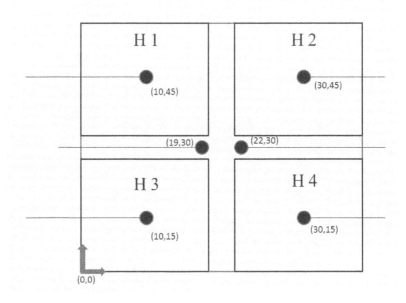

FIGURE 4.3
Thermocouple positions at the heat sink base for discrete heating [65].

FIGURE 4.4
Thermocouple positions at the heat sink base for cylindrical heat sink [62].

measure PCM temperatures on the inside of the heat sink, two thermocouples each are placed at 5, 10 and 15 mm from the base inside the PCM. One thermocouple is placed in still air to measure the ambient temperature. All sixteen thermocouples are bonded to the heat sink using Araldite$^{(TM)}$ epoxy.

4.2 Uncertainty analysis

The uncertainty associated with the measurement of fundamental quantities like the voltage and current can be taken as the least count of the corresponding measurement devices, $\pm$ 0.1V and $\pm$ 0.01A respectively. As far as the temperature measurement is concerned, based on the calibration, it can be inferred that value of temperature is accurate within $\pm 0.2^o$ C. However, the derived quantity power, which is a product of the voltage and current, requires an uncertainty calculation.

The uncertainty in the power can be calculated as

$$\sigma_{p1} = \pm\sqrt{(\frac{\partial P}{\partial V}\sigma_V)^2 + (\frac{\partial P}{\partial I}\sigma_I)^2} \qquad (4.2)$$

For a power level of 1.5 W, the uncertainty in the power turns out to be

$$\sigma_{P1} = \pm\sqrt{(0.31 \times 0.1)^2 + ((4.85 \times 0.01)^2} \qquad (4.3)$$

$$\sigma_{p1} = \pm 0.057W \qquad (4.4)$$

The total power is given by

$$P = P_1 + P_2 + P_3 + P_4 \qquad (4.5)$$

$$\sigma_P = \pm\sqrt{(0.057)^2 \times 4} \qquad (4.6)$$

$$\frac{\sigma_P}{P} = \pm 0.020 \qquad (4.7)$$

$$\sigma_P = \pm 2\% \qquad (4.8)$$

About 10 experiments are randomly chosen and repeated to check the repeatability spread of the data. Since the experiments are conducted in a controlled environment, the ambient temperature remains almost constant during all the experiments. Two parameters are monitored during the repeatability of the experiments. One is the peak temperature after 3000s of heating and the time to reach set point temperature. This is recorded for both the melting and the solidification cycles. The maximum deviation in these quantities is found to be 0.2 K and 22 s respectively. Hence, it is evident that data is highly reliable in terms of thermal performance and optimization.

Before the heat sink is assembled, a leak test is carried out. A fixture which could hold the bare heat sink in still air is used for the leak test. Once the PCM is fully molten, the heat sink is oriented at different angles to check the flow of PCM outside the aluminum wall. Once zero leak is ensured, the heat sink is assembled and the experiments are started.

The heat sink assembly shown in Figure 4.1 is kept inside a large wooden enclosure to ensure that on the outer surfaces of the heat sink, only natural convection heat transfer to the ambient occurs. The dimensions of the wooden enclosure are 1 m × 0.7 m × 0.7 m. Care is also taken to ensure that there are no external disturbances to the flow of heat. A spirit level is used to check the horizontality of the heat sink. AGILENT 34970A data logger is connected to the thermocouples to record the temperatures. A line diagram of the final experimental setup for the 72 pin fin heat sink and matrix heat sink is shown in Figure 4.5.

4.3 Instrumentation for experimentation

4.3.1 Data acquisition system

The data acquisition system as shown in Figure 4.6 employed for the present experimental studies consists of a desktop computer and a data logger. The

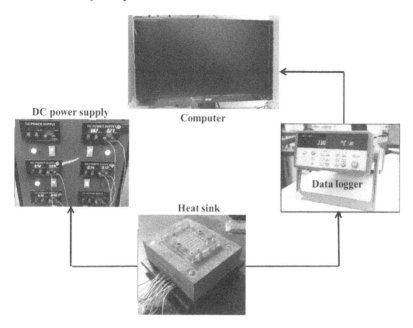

FIGURE 4.5
Line diagram of the arrangement of the experimental setup [65].

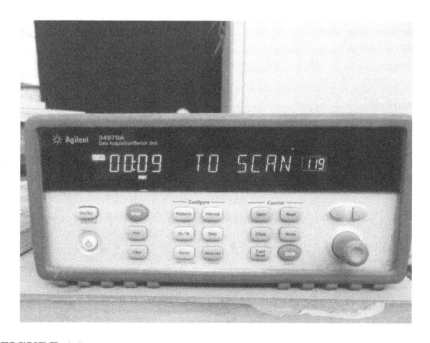

FIGURE 4.6
Photograph of the data logging unit used in the present study.

data logger used is supplied by Agilent Technologies Ltd.(Santa Clara, California, USA), model no. 34970A and the Agilent software([6])is used to record the temperatures, and the data is accessed through a PC-based data acquisition system. The data logger is capable of measuring the temperature from both thermistors and thermocouples.

The data logger can also measure AC and DC voltage, current, frequency, resistance and period. It has three module slots with 20 channels per slot, with each channel connected to a thermocouple. The data logger supports scan rates up to 250 channels per second and has a resolution of 6.5 digits. The data logger is connected to the computer with an RS232 port.

4.3.2 Thermocouples

The thermocouples used in the present study are Chromel-Alumel generally known as "K-type"thermocouples. The thermocouples have an outer yellow Teflon covering containing two single stranded wires. The Chromel wire is coated with a yellow Teflon coating and the Alumel wire has a red Teflon coating. The thermocouples are calibrated against a thermometer of 0.1°C resolution before they are used in the experiments. The calibration is done by immersing the thermometer and the thermocouple in a hot water bath maintained at a fixed temperature. The readings of the thermometer (actual temperature) and the thermocouple (observed temperature) are plotted and fitted with a linear equation. The slope of the equation is the gain and the intercept is the offset.

The calibration equation is $T_{actual} = 1.0039\ T_{observed}$ - 0.0037. The offset correction is due to small stray voltages appearing with the thermo-emf along various points of the thermocouple.

4.3.3 Digital multimeter

The accuracy of the voltage and current readings indicated on the panel of the DC power source are checked with a digital multimeter as shown in Figure 4.7

4.3.4 Constant temperature bath

For the purpose of thermocouple calibration, a constant temperature bath (See Figure 4.8) is used.

4.3.5 DC power source

Figure 4.9 shows the photograph of the TDK-Lambda DC power supply used in the present study. The power supply has a least count of 0.1V and 0.01 A.

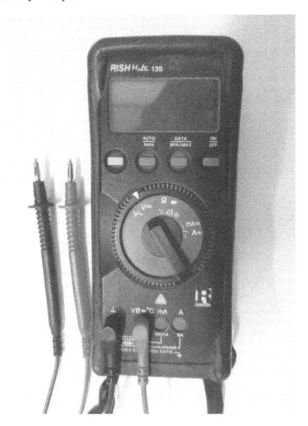

FIGURE 4.7
Photograph of the digital multimeter used in the present study.

4.3.6 Experimental procedure

The procedure followed during experiments is given below.

1. The heat sink assembly, including the plate heater placed in the recess at the bottom of the heat sink is made and kept on a table.

2. The PCM is heated in a borosilicate glass using an electric heater.

3. The liquid PCM is poured into the cavity available in the heat sink and is kept for 12 hours to settle at ambient temperature.

4. The experimental setup is leveled using a spirit level so that the top surface becomes horizontal.

5. The power input is given by switching the DC power supply.

6. At 5 s intervals, temperature data are logged.

7. In order to record the temperature in the solidification phase, DC power source is turned OFF.

8. The procedure is repeated for another power level.

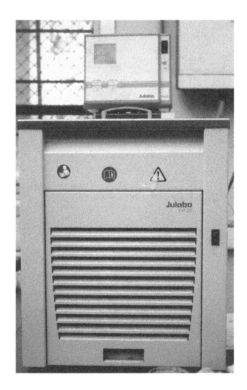

FIGURE 4.8
Photograph of the constant temperature water bath.

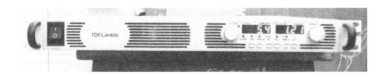

FIGURE 4.9
Photograph of the DC power source used for a single heater.

4.4 Instrumentation for wireless temperature experiments on rotating heat sinks

4.4.1 Wireless temperature measurement module

The CNX t3000 K-Type Wireless Temperature Module as shown in Figure 4.10 provides temperature measurements from −200 to 1372 °C with ± 0.5% accuracy. Additional features of the module include

- 65,000 memory readings

- LCD display

- The module can track up to 10 measurement modules simultaneously, with readings sent to a PC for further analysis

4.4.2 Max31855 amplifier

Thermocouples are very sensitive and require a good amplifier with a cold-compensation reference. The MAX31855 shown in Figure 4.11 is used in this study. This amplifier can be interfaced with any microcontroller, even one without an analog input. This breakout board has the chip itself, a 3.3V

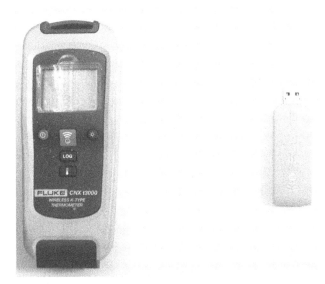

FIGURE 4.10
Photograph of the wireless K type temperature transmitter (left) and receiver (right) module.

FIGURE 4.11
Photograph of the wireless K type thermocouple amplifier.

regulator with 10uF bypass capacitors and level shifting circuitry, all assembled and tested. The advantages of the amplifier are

- Works only with K type thermocouple

- The range of temperature measurements is -200^oC to $+1350^o$ C output in steps of 0.25^o C

- Internal temperature reading

4.4.3   CIC magnetic base angle finder

Since experiments are done at different orientations it is imperative to accurately measure the orientation angle and for this purpose a CIC magnetic base angle finder is used in this study. The key features of this instrument are

- Measures angles as shown in Figure 4.12 accurately and quickly from $0\text{-}90^o$ in any quadrant

- Accuracy within $\frac{1}{2}$ ^{o}

- Magnetic base for mounting

4.4.4   Accelerometer

Since it is important to keep the vibration level within the limit during the rotation of the heat sink, an acceleromenter sensor is employed in this study.

The Dytran model 3053B, shown in Figure 4.13 and used in this work is a rugged, low profile miniature triaxial IEPE accelerometer with a vertical height of 0.9 cm and an overall weight of 6.2 grams. The accelerometer has the following specifications.

- 10 mV/g sensitivity

- 500g range

- 2 to 5,000 Hz frequency range

- 4-pin Dytran designed 1/4-28 radial connector

4.4.5 Fan with heat sink

The specifications of the fan, shown in Figure 4.14, are

- Voltage : 12V DC

- Dimensions : 40 x 40 x 10 mm (LxWxH)

- DC Brushless Motor

4.4.6 Tachometer

The specifications of the tachometer, shown in Figure 4.15 used in the present study are

- Measuring Range 2.5-99999 RPM

- 2.5-19.999 RPM Contact

- Resolution 0.1 RPM, 1 RPM

- Accuracy 0.05

- Sampling Frequency 0.8 s

4.4.7 Lithium polymer battery

A photograph of the Lithium polymer battery used in this study is shown in Figure 4.16. The specifications of the battery are

- Capacity: 1800mAh

- Voltage: 3S1P / 3 Cell / 11.1V

- Discharge: 25C Constant / 50C Burst

- Weight: 161g (including wire, plug and case)

- Dimensions: 115x35x21mm

FIGURE 4.12
Photograph of the angle measurement device.

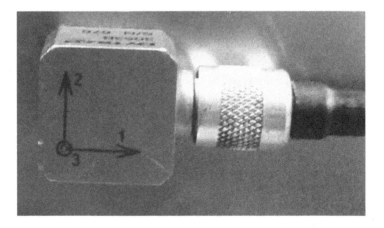

FIGURE 4.13
Photograph of the accelerometer sensor.

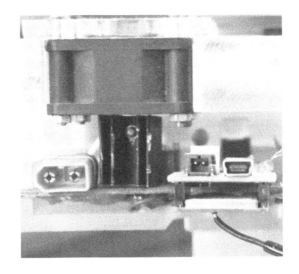

FIGURE 4.14
Photograph of high speed fan-heat sink unit employed to cool the wireless circuit.

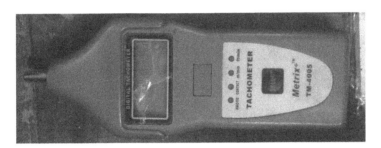

FIGURE 4.15
Photograph of the tachometer used in the present study.

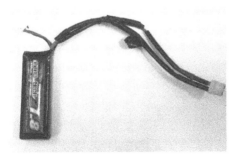

FIGURE 4.16
Photograph of the LiPo battery used in the present study.

4.4.8   Arduino fio

The Arduino Fio used in this study shown in Figure 4.17 is a microcontroller board that runs at 3.3V and 8 MHz. The Arduino Fio is intended for wireless applications. The board has 8 analog inputs, an on-board resonator, a reset button, and holes for mounting pin headers. Connections for a lithium polymer battery and a charge circuit over USB are included in this board. An XBee socket is available on the bottom of the board.

4.4.9   Calibration bath for wireless temperature circuit

A fluke calibration bath as shown in Figure 4.18 is used for the calibration of the wireless temperature circuit. The Fluke bath offers high stability, a large working volume, and flexibility for calibrating a variety of temperature sensors.

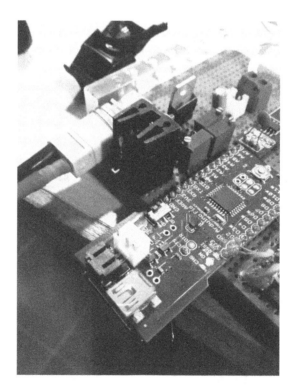

FIGURE 4.17
Photograph of Arduino fio wireless transmitter unit.

FIGURE 4.18
Photograph of the constant temperature bath used for the wireless temperature circuit calibration.

4.4.10   Testing of circuit

The constant current potentiometer circuit is tested in the Robotics laboratory at IIT Madras (see Figure 4.19) with a highly precise voltmeter and ammeter. Both the current and voltage measured by the potentiometer circuit have an uncertainty of less than 2%

4.4.11   Assembled wireless temperature integrated power circuit

Figure 4.20 shows a photograph of the complete wireless temperature and power transmission circuits. The assembled wireless temperature module is integrated with the potentiometer circuit and is assembled at the top of the dynamic rotating structure.

FIGURE 4.19
Photograph of the power regulation being conducted in the laboratory.

FIGURE 4.20
Photograph of the complete wireless temperature and power transmission
circuit.

4.5 Closure

Details of the experimental setup and the instruments used in the present
study were provided in this chapter. The experimental investigation and
results obtained and the discussions thereof for the various PCM-based heat
sinks considered in this study are presented in the subsequent chapters.

5

EXPERIMENTAL INVESTIGATIONS ON 72 PIN FIN HEAT SINK WITH DISCRETE HEATING

5.1 Introduction

In the past, for chip and package thermal modeling and design, thermal engineers used total power dissipation of a chip and a single "junction temperature" to model silicon temperature. Although this approach is still being followed during thermal design for low-power ICs, it is not sufficient for high-performance or power-constrained designs. Increasingly non-uniform power dissipation across the chip leads to local hot spots and elevated temperature gradients across the silicon die. For example, in a 90nm Intel Itanium processor([34]), even after stringent thermal management, local temperature can still be as high as 88°C, while other parts of the die are relatively cool. Therefore, a single value for the total power or the junction temperature without considering the spatial heat distribution is certainly not enough. A thermal package designed for total power and average die temperature inevitably misses the local hot spots, resulting in reliability and performance degradations. In addition, local hot spot temperature is expected to increase as a side-effect of technology scaling. This chapter reports the results of experimental investigations of a phase change material (PCM)-based composite 72 pin fin heat sink subjected to individual heat loading of 4 discrete heaters of equal area. A numerical model is developed to understand the melting heat transfer and the energy storage during melting. This translates to a situation of a single chip with four equal area hotspots of equal and different values of heat dissipation. The ratio of the chip area to the heat sink spreader area is fixed approximately as 1 to neglect the effect of spreading resistance. This study does not include the effect of chip thickness or substrate thickness, which are evidently known to play an important role and may be considered if a detailed numerical model is to be built to study their problems.

5.2   Experimental setup and procedure

The details of the experimental setup for the present study were given in Chapter 4. As mentioned earlier 72 pin fin is proved to be an optimized geometry and hence is chosen for thermal optimization. [52] concluded that an optimum TCE volume fraction of about 10% can give an ideal heat sink performance from the point of view of chip temperature and latent heat storage. Furthermore, 72 pin fins correspond to 10% TCE volume. This value maintained in all the heat sinks (matrix pin fin, cylindrical) is explored in the book.

5.3   Results and discussion

5.3.1   Dimensionless number definition

The following dimensionless numbers are used in this study.
(a) Nusselt number, Nu

In the present study, Nusselt number reflects the time dependent relationship between heat flux q and $\triangle T$. Thus, in the present study the heat flux remains constant with time, whereas $\triangle T$ keeps changing with time. The properties of n-eicosane are the same as used in Baby and Balaji (2013)

$$Nu = \frac{q}{\triangle T}\frac{H}{k} \tag{5.1}$$

(*b*) Fourier number, Fo
The Fourier number depicts dimensionless time and indicates the time for diffusion of heat across the PCM thickness.

$$Fo = \frac{\alpha\ t}{H^2} \tag{5.2}$$

(*c*) Stefan number, Ste
The Stefan number represents the degree of superheating experienced by the liquid PCM. The product of Ste and Fo represents the dimensionless time for phase change without convection.

$$Ste = \frac{C_p\ q_b}{L}\frac{H}{k} \tag{5.3}$$

(*d*) Rayleigh number, Ra
To account for the buoyancy effect inside the liquid PCM that drives the heat transfer, the Rayleigh number, as defined below, is employed

$$Ra = \frac{g\ \beta\ \rho^2\ c_p\ q_b\ H^4}{k^2\ \mu} \tag{5.4}$$

Preliminary experiments were conducted to better understand the effect of different discrete heating situations as listed in Table 5.1. The results discussed in this chapter are from the work [63], [66], [65]. The depiction of the uniform power dissipating chip and a hotspot is as shown in Figure 5.1 Intuitively it is evident that adding more fins on the region of the hottest spot is beneficial as the fins can move the heat more efficiently to the far away region of PCM, but this study assumes that we want to optimize the hotspot location or the magnitude for a fixed heat sink geometry.

TABLE 5.1
Details of the cases investigated

Sl. No	Case	Description
1	A	Single plate heater with heat input of 4W(with PCM)
2	B	Discrete plate heater with uniform heat input of 1W each(with PCM)
3	C	Discrete plate heater with uniform heat input of 1W each (without PCM)
4	D	Discrete plate heater with diagonal heating of 2W each(with PCM)
5	E	Discrete plate heater with H1off H3-2W, H4-1W, H2-1W(with PCM)

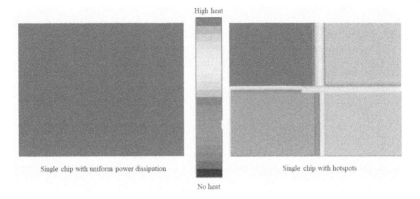

High heat

Single chip with uniform power dissipation   Single chip with hotspots

No heat

FIGURE 5.1
Depiction of hotspot on a single chip.

5.3.2 Effect of uniform heating on the thermal performance of heat sink

Uniform heating here means a supply of equal heat input to all the discrete plate heaters. In this study each discrete plate heater is supplied with an equal heat input of 1W. Baseline comparisons are carried out between case A and case B as shown in Figure 5.2.

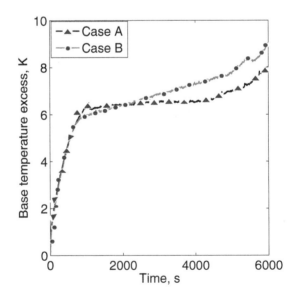

FIGURE 5.2
Comparison of transient temperature between cases A and B (Refer to Table 5.1).

TABLE 5.2
Details of the heat flux in various cases studied [63]

SI. No	Case	Total heat input (W)	Heat flux (W/m2)
1	A	4	1587.3
2	B	4	1724.1
3	D	4	3448.3
4	E	4	2298.9

In both the cases while the heat transfer rate remains the same, heat flux is higher for case B. Table 5.2 shows a quick overview of the operating conditions and the heat fluxes for the various cases analyzed.

Figure 5.2 shows the temperature time history for cases A and B. From Figure 5.2, it is seen that after a charging period of 6000s, the peak temperature reached in case B is lower than case A by 1.5^o C. Hence, as expected in case of discrete plate heaters, the device operating temperature is slightly lower than in a single plate heater case even though the total heat load for case A and B is the same. This is a result of more uniform melting of PCM inside the heat sink. In the case of discrete plate heaters supplied with uniform heat input, the heat from all the four plate heaters aids the melting of PCM uniformly and hence the device temperature does not increase without complete melting of PCM. Even so, the difference in temperature excess even at the end of 6000s is not substantial.

5.3.3 Enhancement in the thermal performance due to PCM

In order to study the enhancement afforded by the use of PCM, a comparison of the performance was carried out (between cases B and C) and the results are shown in Figure 5.3.

From this figure, it is seen that excess temperature after 4800s of heating is 12° C lower in case B, as opposed to case C, which is very significant. When the heat sink is not filled with PCM, the base temperature quickly rises which is highly undesirable. In both the comparisons, it can be seen that in case B the first 1000s is marked by a sensible heating phase. Once the PCM starts melting near the base it can be noted that as the heating continues, the temperature is almost constant with a temperature excess of $6.5 ^o$ C from 1000s to 4800s. The latent heat phase is the more critical one for the safe operation of portable electronic devices. Since the latent heat phase time is stretched, the device can be operated for a longer time under safe conditions. To examine the effect of PCM in the heat sink a new term called enhancement ratio is introduced. The enhancement ratio is defined as the ratio of time taken to reach a particular set point temperature with PCM to that of without PCM. The time taken to reach the set point temperature for case B is 5635s, while that for case C is 570s, thereby leading to an enhancement ratio of 9.6 with the use of PCM in the heat sink.

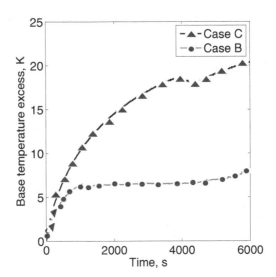

FIGURE 5.3
Comparison of transient temperature between case B and case C(Refer to Table. 5.1)[63].

5.3.4 Effect of diagonal heating on the thermal performance of heat sink

The effect of diagonal heating was examined next. The diagonal heating term (case D) refers to heat input to any one of the diagonal pairs (either (H1, H4) or (H2, H3)) while the other pair is being given zero heat input. Figure 5.4 shows the transient temperature distribution for diagonal heating of discrete plate heaters against a uniform heating of discrete plate heaters at a power level of 4W.

It can be observed from Figure 5.4 that the peak temperature reached after 4800s of heating is 1.5^o C higher for case D than case B. The result indicates that diagonal heating results in increase in the $\triangle$ by 1.5^oC. However, the heat flux in case of diagonal heating is almost twice the heat flux in the case of uniform heating. So, it can be inferred that for a quick estimation of operating time of the device either by analysis or through experiments, the assumption of a single uniform plate heater will not lead to unrealistic results.

5.3.5 Effect of non-uniform heating on the thermal performance of heat sink

Non-uniform heating here means one discrete plate heater is switched off and the other three plate heaters are supplied with heat input which may be uniform or non-uniform. In this study, one heater plate is switched off and

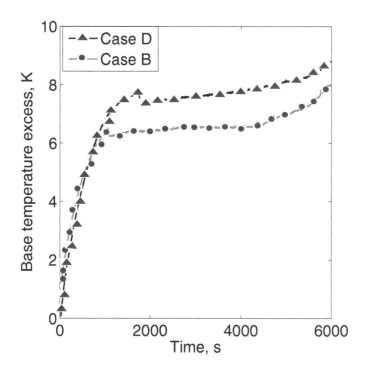

FIGURE 5.4
Baseline comparison of transient temperature distribution between cases B and D(Refer to Table. 5.1) [63].

care is taken to ensure that sum of the heat input to all other discrete plate heaters is 4W. Figure 5.5 shows a comparison of the performance of cases E and B.

From Figure 5.5 it is clear that when one heater is switched off the heat flux has increased and at the end of 4800s of heating, the maximum temperature reached is 2.5° C higher in case E as opposed by case B. While this corroborates one's intuitive reasoning that uniform heating will result in a decreased $\triangle$T, the error involved in using the uniform heating assumption for a discrete heat case will not lead to a significant error in $\triangle$T.

5.3.6 Thermal performance of heat sink without PCM

Investigations on the thermal performance of heat sink without PCM for various power levels are also done in order to appreciate the enhancement in thermal performance afforded by the PCM. In this case, there is no PCM in the heat sink which is heated from the bottom using the discrete plate heaters with uniform power input of 1, 2 and 4W on each plate heater. Figure 5.6

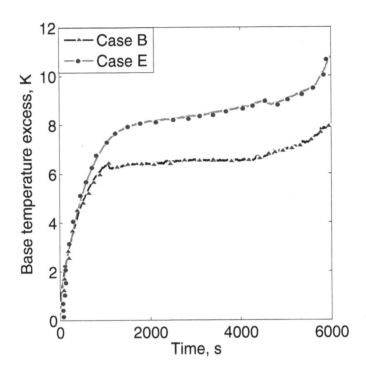

FIGURE 5.5
Comparison of temperature time history for cases B and E (Refer to Table. 5.1)[63].

shows the transient temperature history of the heat sink without PCM for various heat inputs on discrete plate heaters. At higher power inputs, the set point temperature is reached very early, forcing the designer to use a PCM-based heat sink. It was observed that the peak temperature with PCM is lower by more than 10^oC as opposed to the no PCM case, which signifies a highly improved thermal control with the PCM.

5.3.7 Effect of discrete heat source on time taken to reach set point temperature

For a total heat input of 4W, the time taken to reach a set point temperature of $T_\infty + 7.5^oC$ for various cases is presented in Table 5.3. As discussed earlier electronic devices in general have to be operated within a particular set point temperature which is generally the safe operation limit. Table 5.3 shows a comparison of various cases analyzed and the corresponding time taken in each case to reach the set point temperature of 42.5^o C. As expected, a uniformly distributed heat load performs remarkably well compared to other cases.

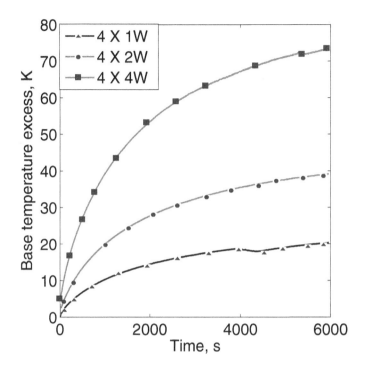

FIGURE 5.6
Comparison of temperature time history for cases without PCM [63].

Furthermore, it is seen that a discrete heat source can stretch the time of operation of the electronic device. Further, it can be seen that case C (which corresponds to 4 heaters of 1W each without PCM) reaches the set point temperature in 570s, which is the least. This is a good indication that PCM helps to maintain the device temperature under a safe operating limit. Diagonal

TABLE 5.3
Time to reach setpoint for various dicrete heating situations (all with PCMs except for case C) [63]

Sl. No	CASE	Time (t), s
1	A	4440
2	B	5635
3	C	570
4	D	2290
5	E	1170

heating of 2 x 2W appears to be much better than 0,1,2,1 W configuration. Case B with uniform distribution of heat load also gives a marginal reduction in $\triangle T$ of 1.5° C against the single plate heater case A after 6000s of heating.

The effect of non-uniform heating when all the heaters are operated simultaneously is now studied. The total power in all the experiments remains 6 W. Throughout the study, the minimum heating value of any heater is kept as 10% of the total heat input and maximum as 70%. All the four heaters are operated simultaneously as already mentioned. To evaluate the performance of the heat sink, the temperature time history of the heat sink base is monitored. Results for the randomly chosen 20 experiments are shown in Table 6.2

In the current study, the performance of the heat sink is defined in terms of time to reach set point temperature as already mentioned. The time to reach

TABLE 5.4
Time to reach set point temperatures for different combinations of heat input for the 6W discrete heating case [65]

Exp ID	Q_1 (W)	Q_2 (W)	Q_3 (W)	Q_4 (W)	t_c (s)	t_d (s)
1	1.5	1.5	1.5	1.5	5830[++]	13145
2	1.8	3	0.6	0.6	4880	12395
3	4.2	0.6	0.6	0.6	4705	15305
4	1.5	1.8	1.2	1.5	4550	11430[++]
5	0.6	2.4	2.4	0.6	5215	16035
6	1	1	2	2	4660	20875
7	0.6	3.6	0.6	1.2	5255	17860
8	3	1	1	1	4200	11945
9	0.6	2.7	0.6	2.1	4970	16905
10	1	1	3	1	4680	11700
11	2.1	0.9	0.6	2.4	4415	20475
12	0.6	3.3	1.5	0.6	4865	25000
13	1.2	1.8	2.4	0.6	4505	23315
14	1.2	1.2	1.2	2.4	4610	26000
15	0.6	1.5	1.8	2.1	4455	25050
16	1.68	0.96	1.44	1.92	4350	28000
17	0.84	1.26	1.68	2.22	4415	19850
18	3.3	0.9	0.6	1.2	4475	28500
19	0.72	1.44	2.16	1.68	4620	21500
20	1.5	0.75	2.25	1.5	4130[--]	27500
21	0.72	1.92	0.96	2.4	4260	29000[--]

set point temperature is one of the key figures of merit used consistently in this study hereinafter. The set point temperature is generally 15 °C above the ambient temperature during the melting cycle and the ambient temperature itself for the solidification cycle. The goals in the design of a heat sink are a longer time to reach set point temperature during charging and a shorter time during the discharge cycle. However, in practice the opposite trend is invariably seen (In practical applications the PCM melts fast and solidifies slowly, which is not preferred. Preliminary results showed that the solidification time can be 4 times higher than the melting time.)

From Table 6.2, it can be seen that uniform heating (1.5, 1.5, 1.5, 1.5) as expected leads to the highest time to reach the set point temperature during the melting process. Additionally, when non-uniform heating situations are considered, if any two heaters receive 10% of the total heat, such an arrangement performs better in the melting or charging phase.

A very interesting observation that can be made from the experimental results (Figure 5.7) is that the cases with any two heaters receiving 10% of the total wattage each shows better performance and lies in Region II in terms of time to reach set point. This is seen to be true for both the 4 and 6 W cases. It is easy to conclude that the remaining two heaters shall be energized with equal heat input to get better performance, but the experiments spring a surprise and it is seen that when the other two heaters receive a heat input of

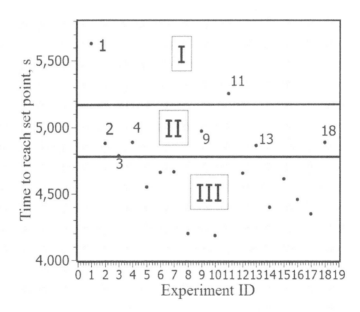

FIGURE 5.7
Variation of time to reach set point temperature for various cases considered in the study [65].

65% and 15% of the total wattage (Case 11 in Table 6.2 and Figure 5.7) each, the heat transfer performance is close to that of the uniform heating case. It is instructive to mention here that the ideal situation of uniform heating on all heaters is rarely encountered in practice.

5.4   Heat transfer correlations

In the present study, for a total wattage of 4 and 6W, a total of 110 experiments are performed and the time to reach setpoint is monitored. (Table 6.2 is only a representative sample)

Table 6.2 shows the time to reach set point for the randomly chosen 21 experiments. Using all of the experimental data, a correlation is obtained for the dimensionless time (Fo) against dimensionless heat inputs(q_n). The quantity q_n is defined as

$$q_n^* = \frac{q_n}{Q}$$

To include the effect of natural convection, the Rayleigh number, Ra is included in the correlation, in such a way that

$$\frac{Fo}{Ra^{0.25}} = aSte^b(q_1^*)^c(q_2^*)^d(q_3^*)^e \tag{5.5}$$

From a multiple linear regression using data fit software, the constant a and exponents b, c, d, e turn out be 0.06, -1.06, 0.024, 0.026 and -0.018 respectively. The correlation in Equation 5.6 can then be rewritten as

$$\frac{Fo}{Ra^{0.25}} = 0.06Ste^{-1.06}(q_1^*)^{0.024}(q_2^*)^{0.026}(q_3^*)^{-0.018} \tag{5.6}$$

The Stefan number exponent can be approximated thus,

$$Ste^{-1.06} \approx Ste^{-1} \tag{5.7}$$

Equation 5.6 on slight readjustment of values a, c, d and e values then becomes

$$\frac{Fo}{Ste^{-1}Ra^{0.25}} = 0.06(q_1^*)^{0.025}(q_2^*)^{0.026}(q_3^*)^{-0.017} \tag{5.8}$$

$$\frac{Fo.Ste}{Ra^{0.25}} = 0.06(q_1^*)^{0.025}(q_2^*)^{0.026}(q_3^*)^{-0.017}(r^2 = 0.958) \tag{5.9}$$

Equation 5.9 has an R^2 of 0.958 and RMS error of 0.009

The term on the left hand side of the correlation is the characteristic time for phase change with natural convection([31]).

$$\tau_{fit} = 0.06(q_1^*)^{0.024}(q_2^*)^{0.026}(q_3^*)^{-0.018} \tag{5.10}$$

Equation 5.10 is valid for the following range of parameters

$$1.27 \leq Ste \leq 1.92$$

$$1.8 \times 10^6 \leq Ra \leq 2.8 \times 10^6$$

In Equations 5.9 and 5.10, q_4 is not included in the correlation explicitly since if q_1 q_2 q_3 are specified q_4 is automatically fixed to satisfy the constraint considered in the study.

5.5   Engineering usefulness of the correlation

Consider the two cases 1 and 2 involving a total power level of 4 and 6 W respectively. Mathematically these can be represented as
   Case 1

$$\sum q_n = 4W \tag{5.11}$$

Case 2

$$\sum q_n = 6W \tag{5.12}$$

In order to avoid local overheating, an individual heater is assigned a minimum of 10% of the total heat input and a maximum of 70% of the total heat input. This can be expressed mathematically as

$$0.1 \leq q_n^* \leq 0.7, \forall n = 1, 2, 3, 4 \tag{5.13}$$

Equation 5.10 can be further simplified as

$$\tau = 0.06 \times \bar{q} \tag{5.14}$$

$$where, \bar{q} = (q_1^*)^{0.024}(q_2^*)^{0.026}(q_3^*)^{-0.018} \tag{5.15}$$

For the uniform heating case

$$(q_1^*) = (q_2^*) = (q_3^*) = 0.25 \tag{5.16}$$

In this case $\bar{q}$ becomes 0.96 which is maximum and translates to the case of uniform heating with a melting time of 5830s. Now, when any two heaters take 10% of the total heat input then the $\bar{q}$ is in the range of 0.91 to 0.93, which corresponds to a time of 5255s which is close to the best performance. The case with any two heaters receiving 10% of the total heat input and the other two receiving 65% and 15% of the heat input is studied with the help of the correlation developed. Upon substitution in Equation 5.10, $\bar{q}$ turns out to be 0.95. This value is very close to the uniform heating case 5.10.

The preceding arguments clearly show that the negative exponent on one of the heat input ratios provides a damping effect on the characteristic time τ. It can also be stated that the value of τ does not increase with increase in power level of all the four heaters.

Additional experiments are conducted for the case of 5W and these points are also added in a parity plot and experimental data corresponding to 5W are indicated as '*' on the plot. It is seen that even for data not included in the development of the correlation the latter performs very well, buttressing the adequacy of the correlation and the results show good agreement, as seen in Figure 5.8

Thus an approximate optimum can be searched in the reduced search space of Q_1 and Q_2. Preliminary experiments indicated that, the situation is not the same when the solidification cycle is also considered. There are better configurations in terms of the time to reach set point than the uniform heating case, which is counter-intuitive and this warrants a detailed investigation. Hence, for more reasons than one, there is a strong need to get to the bottom

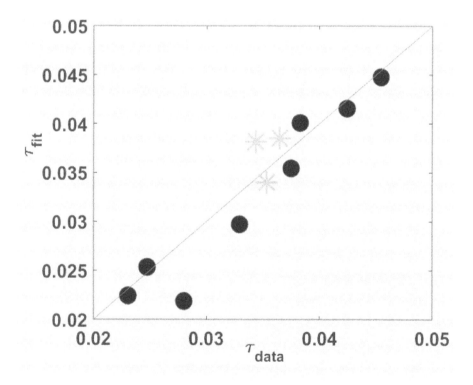

FIGURE 5.8

Parity plot showing agreement between correlation and the experimental data [66].

of the problem and systematically investigate the differential heating of power sources. The subsequent investigations focus on discrete non-uniform heating where $\sum Q_i = 6W$

5.5.1 Performance of diagonal and planar heating at the base

The term diagonal heating refers to a situation where the diagonally opposite heaters, either H_1 and H_4 or H_2 and H_3 in Figure 4.2 operate with maximum wattage. Planar heating refers to the situation where adjacent heaters, either H_1 and H_3 or H_2 and H_4 operate at the maximum wattage. In the present study, with the help of experiments it was evident that planar heating and diagonal heating have the same effect on the melting process. To conclude, what really contributes is the different combination of power inputs or rather the means of distribution of power within the four heaters.

5.5.2 Comparison of uniform heating vs. non-uniform heating at the base

In this study, the sensible heating time is defined as the time taken for the heat sink to heat from ambient temperature to the melting temperature of PCM. The latent heat time is defined as the time in which a heat sink operates in the temperature range of T_{melt} to $T_{melt} + 2^o$ C.

Case 1 refers to uniform heating and case 8 refers to a situation where one heater is energized with 3 W and the other three heaters with 1 W each (shown in Table 6.2).

Table 5.5 shows the sensible heat time (pre-melting) and the latent heat time for the four cases.

From Table 5.5, it is seen that case 1 has the highest sensible heat time and latent heat time. However, case 8 has the lowest of both the sensible and latent heat times, signifying poor heat transfer performance. The ratio of latent heat time to the sensible heating time remains the same for all the four cases. The merit of any PCM-based heat sink is the latent heat time and sensible heating time (before melting). Any optimization study aims to arrive at a configuration of heat sink which has the maximum of the sensible and

TABLE 5.5
Details of sensible heating time and latent heat time for four random cases from Table 6.2 [65]

Exp ID	t_{se}, s	t_L, s	t_L/t_{se}
1	1670	2525	1.51
8	1050	1565	1.5
7	1515	2575	1.51
15	1150	1740	1.51

latent heat time. The convection heat transfer driving the melting process has a direct effect on the base and the wall temperatures.

The density and the dynamic viscosity of the PCM are taken at 309.7K, which corresponds to the melting temperature of the PCM. From the experiments, it is observed that during the melting process, the heat is transferred to the PCM through the following ways

- Directly from the base

- From the heat sink wall

- From the fin surface

Once the PCM near the aluminum surface melts, it forms a self-insulating layer as the thermal conductivity of the PCM in the liquid state is very low (0.15 W/mK). The role of the fins is to reduce the temperature gradient in the composite heat sink due to the "self insulating" effect and promote isothermal phase change process.

The diversity in thermal performance is clearly seen with respect to both melting and solidification. In the case of uniform melting, all the four regions melt uniformly. All the four heaters attain the same temperature at any instant of time. However, this is not the case when a non-uniform heating situation is considered. The transient temperature histories of the four heaters are significantly different at any given instant of time. As observed from the experiments, when one among the four regions is completely molten, the convection current becomes so strong that it erodes the nearby molten PCM layers, thereby increasing the overall base temperature. As soon as the energy input to the heat sink is made zero (heater switched off) the PCM close to the aluminum surface gets solidified first. During the melting process, the heat transfer occurs both by conduction and convection. However, during solidification, the only mode of heat transfer is conduction (considering isothermal phase change). For n-eicosane, the solid phase thermal conductivity is higher than the liquid phase thermal conductivity. At the end of the melting process, the uniform heating case has 100% liquid PCM. However for case 2 in Table 5.4, at the end of the melting process there is partially unmelted PCM in the regions that received low heat input values. Consider the case receiving (3,1,1,1) W of power on the four heaters. During the initial stage of the melting process the Nusselt number at the heated base drops steeply as shown in Figure 5.9.

This is a characteristic of transient heat conduction ([31]). When Fo.Ste >1, the rate of Nusselt number decreases, indicating the onset of convection heat transfer. After Fo.Ste >1.75 the Nusselt number (Nu) approaches an asymptotic value very close to the set point temperature indicating full convection(when the ratio of convection heat transfer to the conduction heat transfer approaches a value much greater than 1), when the heat sink is predominantly occupied by liquid PCM.). The heater is switched off at Fo.Ste = 2.8. At Fo.Ste=2.8, the solidification begins. As can be seen from Figure 5.9,

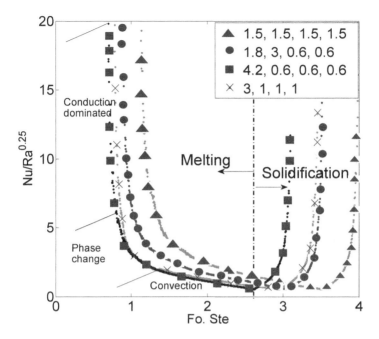

FIGURE 5.9
Variation of scaled heat transfer coeffcient with characteristic time for phase change (The legends are in the form Q_1, Q_2, Q_3, Q_4)[65].

beyond Fo.Ste=2.85, the heat transfer is dominated by conduction. This diversity in melting and solidification phenomena underscores the need for multi-objective optimization.

5.6   Heat loss during experiments

To account for the heat loss that occurs during the experiments, a numerical model of the experiments with the enthalpy-porosity technique was developed and solved with ANSYS Fluent 14.0. The methodology and governing equations of this particular computational technique have been extensively reported in the literature. For the sake of brevity, only the results obtained from the simulations are discussed here. Experiments are conducted to validate the simulations. A wind tunnel as shown in Figure 5.10 is chosen for this purpose. The horizontal wind tunnel consists of a diffuser, settling chamber, nozzle, and rectangular section, which are assembled by using air tight gaskets, and placed on a mild steel stand. Air is supplied by a blower at various velocities (0.01 to 2.5 m/s) to the diffuser through a connecting pipe having a

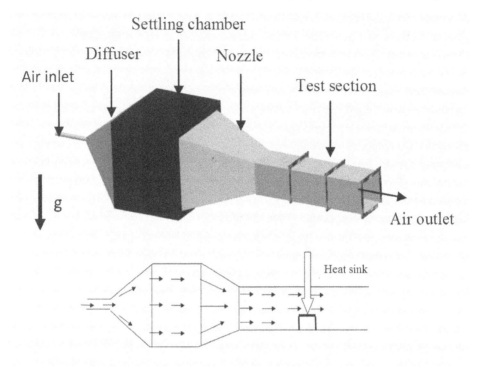

FIGURE 5.10
(a)Schematic of three-dimensional view of the horizontal wind tunnel
employed in this study. (b)Schematic of front view of the horizontal wind
tunnel employed in this study.

diameter of 11 cm and a length 70 cm. The required fluid flow is achieved by
an electronic controller. The diffuser is made up of a mild steel sheet having a
thickness of 5 mm, diameter of 10 cm at the inlet and a square cross-section
$40 \times 40\ cm^2$ at the outlet. From the diffuser, the air flows to the settling chamber of size $40 \times 40\ cm^2$. The settling chamber comprises honeycomb structures
and two wire meshes. The honeycomb structure is made up of around 6500
drinking straws which are glued to each other using an adhesive and placed
in a mild steel frame. Each straw has a diameter of 5 mm and a length of
30 mm in the direction of the fluid flow. Two stainless steel wire meshes are
placed at a distance of 150 and 250 mm from one end of the settling chamber.
The honeycomb structure and the two wire meshes are placed in a settling
chamber for flow straightening and minimizing the flow fluctuations. From the
settling chamber, the air passes through the contraction cone to the entry section and test section and then leaves from the exit section to the atmosphere.
The cross-section of the rectangular channel is $175 \times 175\ mm^2$. The entry sec-

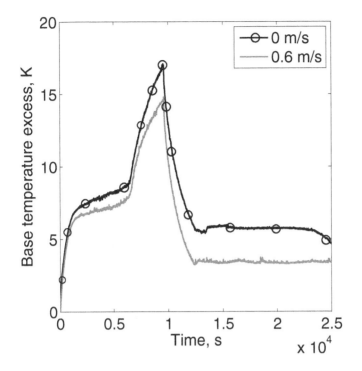

FIGURE 5.11
Effect of external heat transfer coefficient on melting and solidification.

tion which is connected between the contraction cone end to the test section is made of glass having a length of 500 mm. The test section is made of glass walls on three sides and the fourth side is made of removable Hylam plate. This is replaced by either the Teflon block or the flat plate assembly depending on the problem under consideration. The length of the test section is 500 mm. A removable screw is fixed at the entrance of the test section to measure the flow velocity through this small opening using a hot wire anemometer.

The baseline is fixed as free convection. Charging in this process refers to the heating of the PCM-based heat sink using a heater placed at the bottom of the heat sink using a constant DC power supply. Discharging in this process refers to switching of the DC power supply and allowing the PCM to lose the stored heat to the ambient air through the top cover. Figure 5.11 shows a comparison of the performance of the heat sink with 0.6 m/s air velocity over it. From Figure 5.11 it can be observed after a charging period of 9600s the peak temperature reached in the case of 0.6m/s is less than the zero velocity case as it should be.

Forced convection helps to keep the heat sink base at lower temperatures even after 9600s of charging. An external heat transfer coefficient helps satisfy both the objectives simultaneously. It is evident from the experiments that forced convection helps increase the charging time and decrease the discharging time as shown in Figure 5.12.

The time taken to discharge heat considerably reduces as the air velocity is increased from 0 to 0.5m/s. The main goal of the optimization objective in the cooling process using PCMs would be to increase the charging time. In effect, as the charging time is increased the time of operation of the electronic device under prescribed upper temperature limit also increases. An equally important goal is to decrease the discharging time.

Figure 5.12 shows the enhancement in the discharging time for 6W power due to forced convection. The ordinate Y is taken as the scaled time to reach setpoint (Ratio of time taken to reach setpoint to maximum time taken to reach setpoint). Up to 0.3m/s the enhancement is only 3365s. But as the velocity is increased to 0.5m/s the enhancement is 13530s. The discharging time is halved as the velocity is raised from 0 to 0.5m/s. Experiments are conducted for different air velocities on different power levels. Upon analyzing the results, it is seen that the forced convection helps achieve satisfactory

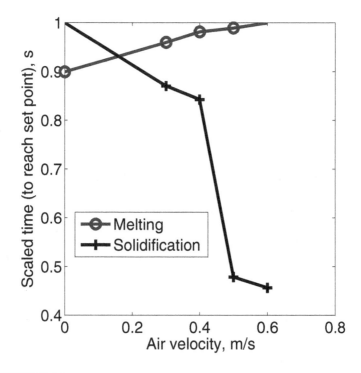

FIGURE 5.12
Enhancement in thermal performance due to increase in external heat transfer coefficient.

improvement during both the charging and discharging cycles. The peak temperature reached at the end of 160 mins decreases as the velocity is increased. It can also be stated that the time to reach a set point temperature increases as the velocity is increased. Furthermore, in the discharging cycle the time taken for the PCM to return to its initial state decreases rather non-linearly as the velocity is increased. As observed for a 4W power level when the velocity is increased from 0 to 0.6 m/s the peak temperature reached after 160mins of heating decreases by 3° C. During the discharging phase the discharge time is nearly halved as the velocity is increased from 0 to 0.5m/s

The numerical model was initially developed with an adiabatic boundary condition on all the outer surfaces, assuming no heat loss. Later the heat transfer coefficient was increased in steps of 0.5 W/m^2K.

For an overall heat transfer coefficient of 1.5 W/m^2K, the numerical results (temperatures at 5 different locations) showed good agreement with the experiments. From further calculations, it was estimated that 0.3 W of heat is lost to the ambient for a total power level of 6W. Therefore, it can be concluded that the heat loss is not expected to play a significant role in further optimization studies.

5.7 Sensible and latent heat accumulation for pin fin heat sink subject to discrete non uniform heating

The mass of the PCM used in the present study is 45g. The latent heat capacity is 234kJ/kg. For 45g of PCM, the latent heat is calculated to be approximately 10.6 kJ. Therefore, for any discrete heating situation, the latent heat remains the same. Natural convection will play a major role in the best utilization of this latent heat. In a latent heat thermal energy storage system, the duration for which the device remains at an almost constant temperature holds the key to a succesful design. An efficiently designed heat sink should make complete use of the latent heat. So to understand the latent heat storage and the sensible heat storage, a numerical model is developed and matched with experiments.

5.7.1 Numerical model

The advantage of using simulations to conduct an extensive parametric study is critical to determine the optimal configuration of the heat sink for maximum performance as detailed parametric study through experimentation is time consuming and expensive.

The modeling of phase-change processes presents a significant challenge due to the complexity of the involved phenomenon ([24]). Some important factors that need to be considered are

- volume expansion during phase change

- convection in liquid phase

- motion of solid in the melt due to density differences

In the current study, the functioning of the heat sink with phase change material is simulated through the ANSYS Fluent 14.0 with the enthalpy porosity formulation developed by [75]. The following assumptions are made during the calculations

- n-eicosane is isotropic and homogenous

- Flow is laminar

- PCM is a Newtonian fluid

- Boussinesq approximation is valid

- Volumetric expansion of the PCM is negligible

The model is meshed using tetrahedral meshing on aluminum material and hexahedral mesh on PCM. The mesh element sizes are 1 and 2mm on the PCM side and aluminum side respectively. The maximum face size and the maximum size of the grid are 1.2 and 2.4mm respectively. The orthogonal quality of the meshing is 0.87.

5.7.2   Governing equations

For the aluminum section, only the conduction equation is considered

$$\frac{\partial}{\partial t}(\rho_s h) = \frac{\partial}{\partial x_i}(k_s \frac{\partial \mathbf{T}}{\partial x_i}) + S \qquad (5.17)$$

In Equation 5.17, h is the sensible enthalpy. The section containing PCM is modeled in three dimensions by considering the flow (in molten state) to be a laminar and incompressible flow.

The continuity equation:

$$\frac{\partial \rho}{\partial t} + \rho(\nabla \cdot \mathbf{v}) = 0 \qquad (5.18)$$

Momentum equations:

$$\frac{\partial \mathbf{v}}{\partial t} + \mathbf{v} \cdot \nabla \mathbf{v} = -\frac{1}{\rho}\nabla p + \nu \nabla^2 \mathbf{v} + \mathbf{g} + S \qquad (5.19)$$

Energy equation:

$$\frac{\partial}{\partial t}(\rho h) + \nabla \cdot (\rho \mathbf{v} h) = \nabla \cdot (k \nabla \mathbf{T}) \qquad (5.20)$$

Melting is taken care of by incorporating the enthalpy porosity formulation in ANSYS Fluent 14.0. The enthalpy h can be written as a sum of sensible enthalpy h_s and the latent heat h_l.

$$h = h_s + h_l \tag{5.21}$$

$$\frac{\partial}{\partial t}(\rho h) + \nabla \cdot (\rho \mathbf{v} h) = \nabla \cdot (k \nabla \mathbf{T}) \tag{5.22}$$

where,

$$h_s = h_{ref} + \int_{T_{ref}}^{T} c_p dT \tag{5.23}$$

The liquid fraction, γ, can be defined following [75] as

$$\gamma = 0 \qquad if \quad T < T_{solidus} \tag{5.24}$$

$$\gamma = 1 \qquad if \quad T > T_{liquidus} \tag{5.25}$$

$$\gamma = \frac{T - T_{solidus}}{T_{liquidus} - T_{solidus}} \qquad if \quad T_{solidus} < T < T_{liquidus} \tag{5.26}$$

The latent heat content can now be written in terms of the latent heat of the material

$$h_l = \gamma L \tag{5.27}$$

The enthalpy-porosity technique treats the mushy region (partially solidified region) as a porous medium ([75]). The porosity in each cell is set equal to the liquid fraction in that cell. In the fully solidified regions, the porosity is equal to zero, which extinguishes the velocities in these regions. Voller and Prakash [75] suggested that the momentum sink due to the reduced porosity in the mushy zone takes the following form

$$S = \frac{(1 - \gamma)^2}{(\gamma^3 + c)} A_{mush} \mathbf{v} \tag{5.28}$$

Here 'c' is a small number (0.001), to prevent the division of the numerator by zero. When heat is added to a fluid and the fluid density varies with temperature, a flow can be induced due to the force of gravity acting on the density variations. Such buoyancy-driven flows are termed natural-convection (or mixed-convection) flows. The modeling is done by involving the Boussinesq approximation. This model treats density as a constant value in all solved equations, except for the buoyancy term in the momentum equation

$$(\rho - \rho_0)g \approx -\rho_0 \beta (T - T_0)g \tag{5.29}$$

The Boussinesq approximation is given as follows

$$\rho = \rho_0(1 - \beta T) \tag{5.30}$$

This effectively eliminates the density ρ from the buoyancy approximation.

FIGURE 5.13
Picture showing the computational mesh employed.

The change in material properties of the material with melting are modeled using a piecewise linear model for the specific heat and thermal conductivity. The heat sink is meshed with tetrahedral meshes on the aluminum material and with hexahedral mesh on the PCM as shown in Figure 5.13. In all, about 2,48,000 nodes are used.

SIMPLE algorithm is used with pressure velocity coupling. PRESTO scheme is adopted for discretizing of pressure and SIMPLE algorithm with the second order upwind scheme used. The convergence is set to $1 \times 10^{-6}, 1 \times 10^{-3}$ and 1×10^{-6} for the continuity, momentum and energy equations respectively. An energy balance accuracy of 4.332×10^{-9} was achieved after convergence. The heat sink is subjected to two cycles viz. heating and cooling. During the heating cycle, the heat sink is subjected to a constant power input of 6W.

The numerical model thus developed is validated with experiments and a good agreement is seen, as shown in Figure 5.14. Heat losses occur around

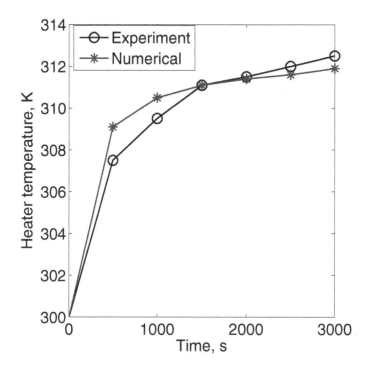

FIGURE 5.14
Comparison of base temperature time history obtained from experiments with the numerical results.

the heat sink owing to natural convection and these need to be quantified. Although the heat transfer coefficient varies with the surface temperature, an overall average heat transfer coefficient is determined by comparing the temperature time histories simulated for various values of U with respect to the experimental temperature time history.

The heat accumulation in a PCM-based heat sink consists of both latent heat and sensible heat

$$Q = Q_L + Q_S \tag{5.31}$$

The latent heat accumulated in the PCM can be estimated by the known melt fraction of the PCM and the enthalpy of melting. The sensible heat accumulated comprises three parts namely

- Sensible heat accumulated in the base and fins

- Sensible heat accumulated in the PCM from its initial temperature to melting

- Sensible heat accumulated in the PCM from melting temperature to its final state

The latent heat is defined as

$$Q_L = m_{pcm} \times L \tag{5.32}$$

For a constant heat flux, sensible heat is defined as

$$Q_S = m \times C_p \frac{q_b}{k} \frac{H}{} \tag{5.33}$$

The calculation of 'm' for PCM is not so straightforward here as four different interacting regimes in the heat sink volume induced by four discrete heaters are present. It is clear that it is not possible to calculate the heat accumulated in the PCM exactly due to the fact that the volumes occupied by the solid and the liquid PCM are irregular. However, the only useful estimate could be from the liquid fraction data. The mass of PCM cannot be divided by 4 for each heater's input. However the analysis can be taken forward with the help of a numerical model. The heat supplied to aid the melting of PCM is from three sources (base, wall and fin surface) and not just from the base.

When discrete heating is done from the base, the whole of the heat sink cavity with fins and PCM is divided into four regions as shown in Figure 5.15, each receiving independent heat input. Each zone comprises three regimes, as shown in Figure 5.15 as follows

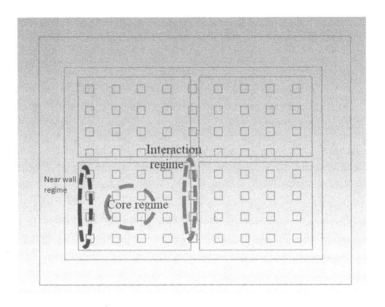

FIGURE 5.15
Picture showing different regimes analysed inside a zone for the 72 pin fin case subject to discrete heating [65].

- Middle or core regime

  The PCM and the TCEs in this regime take up heat directly from the base

- Near wall regime

  The PCM and TCES in this regime take up heat both from the base as well as conducted heat from the side walls

- Interaction regime

  PCM and TCEs in this regime are subjected to heat loading from two or more heaters as they lie at the zone interface. In this regime, a larger gradient of temperature is expected which induces convection.

5.7.3 Uniform heating

At the beginning of the melting process, the total heat input to the heat sink is transferred to the PCM as sensible heat (Figure 5.16).

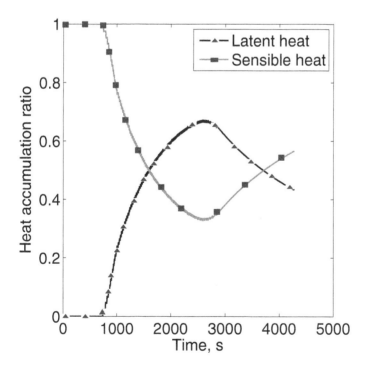

FIGURE 5.16

Latent and sensible heat accumulation for case 1 (Refer to Table 6.2) [65].

The latent heat enters the melting process as time progresses and reaches its maximum share of 63%. Shatikian et al. [61] determined that for a PCM-based heat sink the maximum share of latent heat accumulation does not exceed 70%. As the melting progresses the share of the latent heat drops suddenly due to the 'self-insulation' effect. This gives room for more sensible heat accumulation of the liquid PCM.

5.7.4   Non-uniform heating

For the case of non-uniform heating the four heaters are energized with different heat input. Q_1, Q_2, Q_3 and $Q_4 = 3$, 1, 1 and 1 W respectively. The volume averaged melt fraction values for cases 1 and 8 are shown in Figure 5.17.

Case 8 (3,1,1,1 W) melts 250s early compared to case 1 (1.5,1.5,1.5,1.5 W). This is a direct indication that the liquid phase convection becomes dominant in case 8 earlier than case 1 resulting in quick temperature rise. From further investigation on the liquid PCM velocity for both the cases one can infer

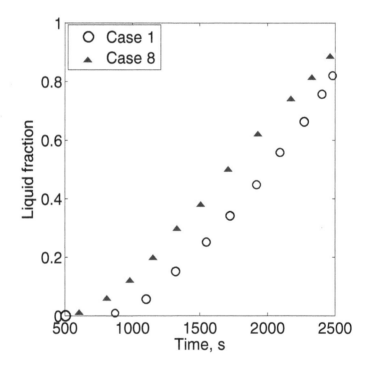

FIGURE 5.17
Transient history of liquid fraction for cases 1 and 8 (Refer to Table 6.2)[65].

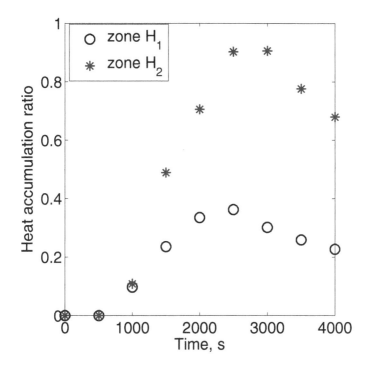

FIGURE 5.18
Latent and sensible heat accumulation for two zones of case 8 [65].

that case 8 presents with high velocity which erodes the nearby solid PCM
at a much higher rate. From Figure 5.18 something really surprising can be
observed. The zone receiving 1W heat has a latent heat accumulation of more
than 70% (high as 90%).

This observation helps one understand the fact that the melting in that zone
is a consequence of the heat not only from heater H_2 but also the convection
heat transfer from the nearby interaction zones. During the melting process,
the increased volume of liquid does not only possess high heat capacity, but
also high thermal resistance for the flow of heat. As the supplied heat remains
the same, the base temperature increases for various discrete heating cases
owing to the effect of high thermal resistance. Natural convection plays a
predominant role in transferring the heat between various regimes inside the
heat sink volume. To understand the melting in each zone, a spherical fluid
volume was created which was centered on the core/middle regime of each
zone and extended around the near wall and interaction regime. The variation
of local melt fraction gives a clear picture of the effect of convection on the
melting process. The PCM in zone H_1 melts 700s earlier than that in zone H_2
(Figure 5.19).

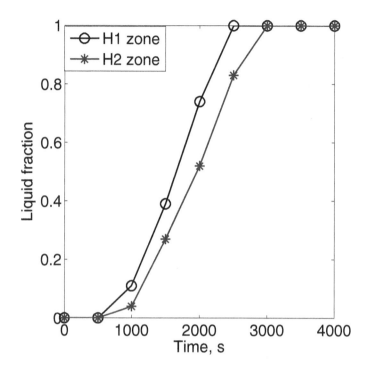

FIGURE 5.19
Time history of liquid fraction for case 8 in zone H_1 (3W) and zone H_2 (1W)
[65].

Analysis of zone H_1 can be done by dividing the zone into three different
regimes, as described earlier. Figure 5.20 shows the liquid fraction time history
in all the three regimes in the zone H_1.

The PCM in the core regime melts faster by taking the heat from the base
directly. Secondly, the PCM in the near wall regime melts at an average pace.
These observations are made at the mid-plane from the base of the heat sink.
The PCM at the interaction regime melts very slowly compared to that in the
other two regimes. The effect of convection is less compared to that of other
regimes and hence a slow rate of melting is seen. Figures 5.21 and 5.22 show
the liquid fraction contours for case 8 for t=500 and 2500s respectively. If there
were no fins to enhance thermal conductivity the PCM in the core regimes
melts slowly. This leads to an increase in the temperature of H_1, thereby
increasing the average base temperature. As soon as the PCM melts in the
zone H_1, the convection currents go on to erode the nearby zone solid PCM.
It is important to understand the variation of liquid PCM velocity inside the
heat sink. Initially, conduction is the dominant mode of heat transfer and
hence zero velocity is encountered. Post conduction dominated melting, very

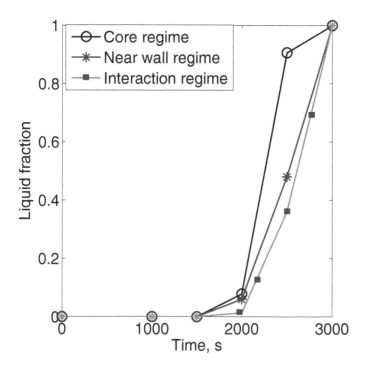

FIGURE 5.20
Transient history of liquid fraction for case 8 zone H_1 in three regimes [65].

low velocity currents begin to form, which is an indication of the onset of convection heat transfer in the melting process. The velocity keeps increasing indicating more convection heat transfer. After this there is a sudden drop in velocity of the liquid PCM.

To reason out the sudden drop in velocity, the vertical temperature gradient was monitored near the fin surface. The velocity of the rotating vortices as shown in Figure 5.23 is induced by the density difference, which is induced by the temperature gradient. As can be seen during the early stages of melting the temperature gradient is negative (higher temperature near the base).

As time progresses the temperature gradient becomes almost zero indicating an isothermal condition as shown in Figure 5.24. After 90% of the PCM has melted the temperature gradient becomes positive (high temperature near the top surface). Hence, the buoyancy source term in the momentum equation becomes negative and leads to a drop in the velcoity. The clockwise rotating vortices change to counter-clockwise rotating vortices as shown in Figure 5.23. So the change from clockwise to anti-clockwise rotation leads to a sudden drop in the liquid PCM velocity. This non-uniform melting is the major disadvantage of non-uniform heating. For the calculation of latent heat

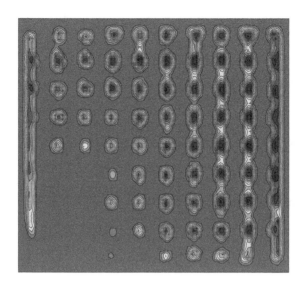

FIGURE 5.21
Liquid fraction contours for case 8 at t=500s [65].

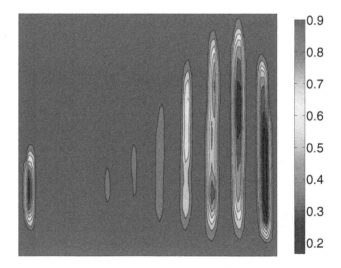

FIGURE 5.22
Liquid fraction contours for case 8 at t=2500s [65].

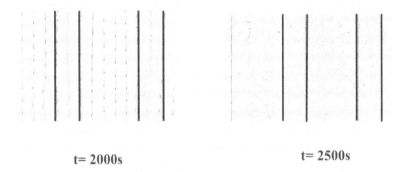

t= 2000s   t= 2500s

FIGURE 5.23
Velocity vector near the fin surface for a 72 pin heat sink under uniform
heating [65].

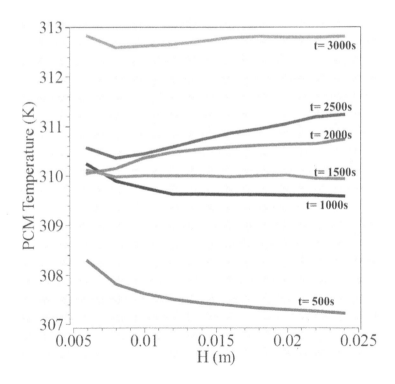

FIGURE 5.24
Variation of near fin temperature at various time instants for a 72 pin fin heat
sink under uniform heating (Refer to Table 6.2).

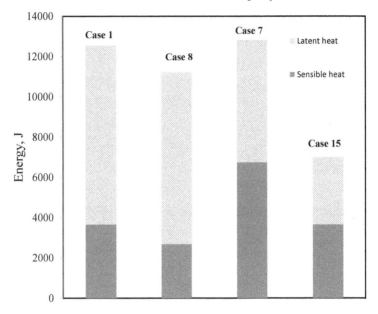

FIGURE 5.25
Latent heat accumulation for cases 1,6 and 8 [65].

stored and sensible heat stored, only the significant regions are taken into account.

Figure 5.25 shows a histogram of latent and sensible heat accumulation for four randomly chosen discrete heating cases. The latent heat share for the various cases helps one understand the performance of the heat sink under different non-uniform heating situations.

5.8   Conclusions

Experiments were performed on a PCM-based 72 pin fin heat sink with 4 discrete heaters at the base. The heat sink was made of aluminum and the PCM used was n-eicosane. Experiments for 110 different combinations of heating level were conducted keeping the total power input equal to 4 and 6W as the case may be. There was a significant diversity observed in the time to reach set point.

The salient conclusions of the study are

- Correlations for the thermal performance (defined as the time to reach set point temperature) of this particular heat sink were developed as a function of the discrete power levels that can accurately predict the performance of the heat sink.

- A heat sink with single plate heater and discrete uniform heating yield the same performance.

- Discrete non-uniform heating has a significant effect on the thermal performance of the heat sink.

- During melting, the ratio of the latent heat time to the sensible heating time remains the same for all the discrete heating cases. During melting, when two of the heaters received 10% of the total wattage, the melting time is maximum.

5.9 Closure

This chapter reported the results of experimental investigations on the effect of discrete heating a pin fin heat sink with PCM which takes into account both the charging and discharging cycles. Experimental results thus obtained can be used to drive even more robust optimization engines to solve multi-objective problems of this class. In the next chapter, candidate multi-objective algorithms are explored and optimal configurations of discrete heating are arrived at.

6

MULTI-OBJECTIVE OPTIMIZATION ALGORITHMS FOR 72 PIN FIN HEAT SINKS

6.1 Introduction

In the previous chapter, the need and motivation for multi-objective thermal optimization of the 72 pin fin heat sink was established. This chapter deals with the performance assessment of distinct (Multi-objective optimization algorithms) MOOs and the experimental validation of Pareto optimal solutions obtained thereof. The results discussed in this chapter are from the work [65].

6.2 Application of multi-objective optimization algorithms

Researchers have been investigating the use of optimization techniques to perform thermo-geometric optimization of composite heat sinks (a PCM-based

89

heat sink with fins). In the last two decades evolutionary multi-objective (EMO) algorithms are being increasingly used. The applications of such algorithms are easily seen in several branches of engineering, particularly in problems with conflicting multiple objectives. The origin of EMOs which mimic evolution dates back to the early 1950s. The use of population during the iterations particularly make the EMOs more suitable for multi-objective optimization as they are capable of finding multiple Pareto optimal solutions in a single run.

From the review of literature presented in Chapter 2, it is very clear that (i) optimization of the discrete heaters in a PCM-based heat sink and (ii) a critical comparison of the candidate multi-objective optimization techniques on the thermal performance of heat sink have not been adequately addressed. In general, multi-objective optimization algorithms are classified into a priori methods, a posteriori methods (evolutionary algorithms) and meta-heuristic swarm intelligence techniques. For this study, one from each of the above mentioned classes of multi-objective techniques is chosen.

1. A priori method - Goal programming

2. Evolutionary algorithm - NSGA-II

3. Swarm intelligence - Particle swarm optimization(PSO)

4. Brute force (For benchmarking)

Additionally, in most of the optimization studies the discharging cycle time has traditionally not been considered as one of the objectives. During the operation of the heat sink, the charging cycle takes place typically for about $1/5^{th}$ of the total operating time, while the discharging cycle is $4/5^{th}$. An important goal of the study is to obtain a set of solutions which can maximize the charging cycle time and minimize the discharging cycle time simultaneously in a PCM-based composite heat sink. This is challenging due to the two objectives being in fundamental conflict with each other.

6.3 Experimental results for 72 pin heat sinks with discrete heating

A detailed description of the experimental setup used in the present study was presented in Chapter 4. The heat sink consists of 72 aluminum pin fins. The dimensions of the heat sink were decided based on the average dimensions of the existing portable electronic devices. The heat sink dimensions are 80 x 62 mm^2 with a height of 25 mm. The heat sink walls are insulated by using cork except for the top portion which is covered with acrylic. The fin dimensions are 2 x 2 x 20 (height) mm^3. Four discrete heat sources are placed at the bottom of the heat sink.

The goal of the experiments is to study the effect of discrete heating on the charging and discharging cycle. Hence the inputs are Q_1, Q_2, Q_3 and Q_4 while the outputs measured are the time taken to reach the set-point (charging time) and the time taken to reach ambient temperature (discharging time). The attainment of set-point and ambient temperature is gauged by the average temperature of the base, which in turn is found to be nearly the same as the temperature measured by the thermocouple located at the centre of the base, in between the four discrete heaters. The linear constraint applied to heat inputs is that the sum of the power of the four discrete heaters must be constant for a given set of experiments. The two heating power levels examined are 4 and 6W.

For both the 4 and 6W total heating power cases, based on preliminary studies it is seen that the case of equal discrete heating, which is equivalent to using one uniform heater, gives the maximum charging time. However, when one considers the discharge time, the best values are not obtained with the uniform heating case. The usefulness of the study stems from the fact that there are only pre-assigned positions to heat sources in a PCB. Studies like this will help the designer place the heat sources (of different strengths) appropriately in these positions to maximize the thermal performance. Typical results of the experiments were already shown in Table 6.2 of Chapter 5.

6.4 Artificial neural network

An Artificial Neural Network (ANN) is a non-linear regression tool inspired by biological neurons. Neural networks are artificial intelligence structures developed from massively interconnected neurons. The operational procedure of a neural network is modeled after the working of the human brain, in which the participation of millions of neurons simultaneously is known to facilitate pattern recognition and complex learning processes. A neural network consists of an input layer, one or more hidden layers and an output layer as shown in Figure 6.1. The hidden layer consists of nodes which assign weights to their inputs and add some biases. These are decided while training the neural network for the given input-output data.

Each input is multiplied by a weight and is then summed up. The summed up value evaluated in the activation function is the network's output. The Levenberg-Marquadt algorithm ([11]) is used for the purpose of regression. The back propagation algorithm is implemented, so as to adjust the weights to minimize the error between the network predicted output and the desired output.

$$n = \sum w_i x_i \tag{6.1}$$

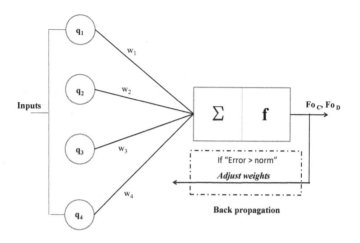

FIGURE 6.1
Schematic of the feed forward back propagation artificial neural network employed in the present study [65].

In Equation 6.1 w_i is the weighting vector.
x_i is the input vector and the output is given by

$$output = \frac{1}{1 + exp(-n)} \tag{6.2}$$

The above activation function produces an output varying between 0 and 1. Before the model is used for predictions, the weight matrix/vector is modified from a random initial state to a fixed equilibrium state. A network's output consists of the values of the weights that reduce the sum of square of the error between the output and the input.

The inputs to the ANN are the non-dimensional heat input values q_1, q_2, q_3, q_4 and the outputs are the non-dimensionalized heating time $(1/Fo_C)$ and the cooling time (Fo_D) represented non-dimensionally as the Fourier number (Fo)

$$q_i = \frac{Q_i}{q_{max}} \quad i = 1, 2, 3, 4 \tag{6.3}$$

$$Fo_c = \frac{\alpha.t_c}{L^2} \tag{6.4}$$

$$Fo_D = \frac{\alpha.t_d}{L^2} \tag{6.5}$$

Since the discrete heating grid offers symmetry, the assumption that symmetric heating inputs would lead to identical outputs in terms of melting and solidification behavior is made use of. Hence, while training the ANN, 'x' experimental data samples can generally be used to give '4x' training points. Therefore, 22 experiments yield 88 training values for the ANN in the 6W case and 12 experiments yield 48 training values for the 4W case. Exploiting the symmetry helps reduce the number of experiments that need to be conducted and also improves the accuracy of the neural network since it learns the symmetry of the grid by itself.

The data from experiments is used as input for generating the neural networks. About 70% of the data is used to train the network and the weights are adjusted according to the errors for this data subset.

About 15% is used to validate the network by measuring network generalization and halting training when the generalization stops improving. The remaining 15% of the data is used for testing the performance of the network both during and after the training. The architecture of the ANN employed in this study is shown in Figure 6.11. MATLAB R2012b Neural Network Fitting application (Demuth 1993) is used to generate the neural network. As shown, the number of neurons in each hidden layer need not be the same and the design of the network itself is a significant part. In general, one hidden layer is sufficient for the large majority of problems. In order to choose the number of neurons in the hidden layers, an iterative procedure is used where the number of neurons in the hidden layer is varied from 1 to 40 and neural networks are generated for each case. The cases are then compared on the basis of the mean relative error, mean square error and coefficient of determination to determine the optimum number of neurons

Matlab ([70]) by default minimizes both the objectives, henceforth $1/Fo_C$ is taken following the duality principle. Before training the data using ANN, an effective sampling technique (Latin hypercube sampling) as described in [70] is used to generate samples that can represent the population sufficiently. The number of variables in this problem is four. The usual practice is to evaluate the model for the input points that are not used in the training process (ANN is basically interpolation). Since 'supervised learning' technique is used here, the inputs to the ANN are accompanied by their desired outputs. When the untrained input points are given as an argument to the network function, a close match is found between the network's output and the desired output. The closeness of the match is indicated by the low value of RMSE (Root Mean Squared error) and MRE (Mean Relative Estimation error).

Hence, the final choice of the network is based on a holistic assessment of several key performance metrics. The results of the neuron independence study are shown in Table 6.1. From Table 6.1, it can be seen that one hidden layer with 32 neurons gives the best performance. Preliminary investigations confirmed that there is little to choose one and two hidden layers. Hence, subsequent optimization studies for this problem were done with 1 hidden layer containing 32 neurons.

TABLE 6.1

Results of the neuron independence study

Number of hidden neurons	MRE	R^2	RMSE
2	0.757	0.885	4.317
4	0.308	0.931	3.337
6	0.346	0.92	3.612
8	0.279	0.937	3.203
10	0.323	0.916	3.694
12	0.332	0.923	3.527
14	0.331	0.924	3.508
16	0.346	0.917	3.667
18	0.307	0.937	3.199
20	0.334	0.928	3.409
22	0.388	0.87	4.582
24	0.361	0.933	3.304
26	0.357	0.898	4.063
28	0.369	0.87	4.592
30	0.428	0.837	5.145
32	**0.264**	**0.951**	**2.814**
34	0.309	0.912	3.77
36	0.386	0.931	3.352
38	0.284	0.933	3.297

6.5 Optimization of discrete heat input of 72 pin fin heat sinks

To better understand the motivation to perform optimization, Table 6.2 from Chapter 5 is reproduced here for the sake of completeness. Experimentally for the 6W total heating power case, one finds that the case of equal discrete heating, which is equivalent to using one uniform heater, gives the maximum charging time. However, when we consider the discharge time, the best experimental values are not obtained with the uniform heating case. When the experimental data is ordered in the increasing order of charging time or decreasing order of discharging time, the order of the discharging time and charging time respectively does not reflect a simple complementary trend, hence paving the way for a requirement of a rigorous multi-objective optimization. The above statements were found to be true for the 4W case too.

The desirable Fo_C (highest value desirable) and Fo_D (lowest value desirable) are indicated by the superscript "$++$" in Table 6.2. Similarly, the most undesirable values are indicated by "$--$". The striking observation one can make is $++$ and "$--$" do not occur together for the same configuration of

TABLE 6.2
Time to reach set point temperatures for different combinations of heat input
for the 6W discrete heating case [65]

Exp ID	Q_1 (W)	Q_2 (W)	Q_3 (W)	Q_4 (W)	t_c (s)	t_d (s)
1	1.5	1.5	1.5	1.5	5830[++]	13145
2	1.8	3	0.6	0.6	4880	12395
3	4.2	0.6	0.6	0.6	4705	15305
4	1.5	1.8	1.2	1.5	4550	11430[++]
5	0.6	2.4	2.4	0.6	5215	16035
6	1	1	2	2	4660	20875
7	0.6	3.6	0.6	1.2	5255	17860
8	3	1	1	1	4200	11945
9	0.6	2.7	0.6	2.1	4970	16905
10	1	1	3	1	4680	11700
11	2.1	0.9	0.6	2.4	4415	20475
12	0.6	3.3	1.5	0.6	4865	25000
13	1.2	1.8	2.4	0.6	4505	23315
14	1.2	1.2	1.2	2.4	4610	26000
15	0.6	1.5	1.8	2.1	4455	25050
16	1.68	0.96	1.44	1.92	4350	28000
17	0.84	1.26	1.68	2.22	4415	19850
18	3.3	0.9	0.6	1.2	4475	28500
19	0.72	1.44	2.16	1.68	4620	21500
20	1.5	0.75	2.25	1.5	4130[− −]	27500
21	0.72	1.92	0.96	2.4	4260	29000[− −]

power level. However, we desire solutions that simultaneously meet both the
objectives. In the present work, four techniques have been employed to determine the set of Pareto optimal solutions (solutions that are not dominated by
any member on the feasible objective function space) which satisfy both the
objectives simultaneously.

1. NSGA-II

2. Goal programming

3. Particle swarm optimization

4. Brute-force search

The performance of the techniques is judged on the basis of the spread of solutions, the proximity to true Pareto front and the computation time required. The metrics observed are

1. Time taken by the algorithm

2. Number of function evaluations

3. Diversity of the solutions

4. Size of Pareto front (number of Pareto front points)

In order to create a robust and accurate optimization framework and thereby evaluate an optimum solution set of discrete heating levels, experimentally generated data are used to train a neural network, which then drives the multi-objective optimization engines. For all the above-mentioned techniques, the fitness function or objective function is derived from the output of the Artificial Neural Network (ANN). The ANN acts as a surrogate for the experiments. This is required in view of the fact that the solution to the optimization problem not only requires values of the output (charging and discharging time) for several combinations of the input (Individual heater values) but also at these values of the input at which experimental results are not available. The feed forward-based back propagation technique was used for the development of the neural network. Details of the ANN employed in this work were elucidated in Section 6.4.

In mathematical terms, a multi-objective optimization problem can be formulated as [70]

$$min(f_1(\mathbf{x}), f_2(\mathbf{x}), ..., f_k(\mathbf{x})) \qquad (6.6)$$

$$g_i(\mathbf{x} \le 0(j = 1, 2, ..., m)) \qquad (6.7)$$

where, $x \in X$ having k objectives and m constraints (by function g).

A solution x^1 is said to Pareto dominate another solution x^2 if

$$f_i(x^1) \le f_i(x^2) \ \forall i \ \in \ (1, 2, ...k) \qquad (6.8)$$

$$f_j(x^1) < f_j(x^2) \ for \ at \ least \ one \ j \in (1, 2, ...k) \qquad (6.9)$$

6.5.1   Latin hypercube sampling

Different combinations of heat inputs are generated using the Latin Hypercube Sampling (LHS) technique ([70]). LHS is a statistical method for generating a sample of the plausible collection of parameter values from a multidimensional distribution. A sampling function of N variables is divided into M intervals. M sample points are then placed to satisfy the Latin Hypercube requirements. In the present study, the total heat input is kept constant while the individual wattages can vary (N=3) with the fourth heat input getting fixed automatically by the constraint of a fixed total power. M corresponds to the number of

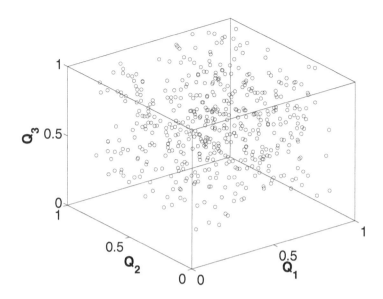

FIGURE 6.2
Samples generated by Latin hypercube sampling technique.

TABLE 6.3
Description of the design variables and the bounds

Name	Feasible range	Type
Q_1	[0.6,4.2]	continuous
Q_2	[0.6,4.2]	continuous
Q_3	[0.6,4.2]	continuous
Q_4	[0.6,4.2]	continuous

combination generated which is 40 in this study. The samples generated are shown in Figure 6.2.

The bounds of the design variables and parameters are shown in Table 6.3.

6.6 Goal programming

Goal programming ([14]) is an extension of the well known linear programming technique ([14]). Goal programming is posed as a non-linear programming problem and takes into account multiple objectives, while linear programming can handle only one objective. The foremost objective of goal programming is to determine a solution that satisfies the constraints and comes closest to meeting the stated goals. In this study, goal programming is used to obtain a set of non-dominated solutions for the two conflicting objectives of maximizing the charging time and minimizing the discharging time.

6.6.1   Problem formulation

$$goal\ 1\ (f1 \leq minimum(f1)) \qquad\qquad (6.10)$$

$$goal\ 2\ (f2 \leq minimum(f2)) \qquad\qquad (6.11)$$

subject to

$$x_1 + x_2 + x_3 + x_4 = 6 \qquad\qquad (6.12)$$

$$0.6 \leq x_i \leq 4.2, \forall\ i\ = 1, 2, 3, 4 \qquad\qquad (6.13)$$

The goal here is the Utopian solution. A utopian solution is one solution that is optimal for all the objective functions that are considered. For a single objective optimization study, the global optimum is the Utopian solution. For a multi-objective optimization problem where conflicting objectives exist, the Utopian solution cannot be obtained. Utopian solution in this study refers to the positive ideal solution, which in this case is[0.4881, 1.812]. The above stated problem is now reformulated in Matlab R14 as

$$Minimize\ p \qquad\qquad (6.14)$$

where 'p' is the deviation from the stated goal

$$subject\ to\ f_i - w_i p \leq goal \qquad\qquad (6.15)$$

The steps involved to obtain the Pareto optimal solution is as follows

- min f_1 -> keeping f_2 constant
- min f_2 -> keeping f_1 constant
- Set goal = [min f_1, minf_2]
- Set initial guess = [x_1 x_2 x_3 x_4]
- Vary the weights from 0 to 1

The main advantage of the goal programming method is that the optimization problem can be posed as a Non-Linear programming (NLP) problem, which is solved by using sequential quadratic programming technique. The Sequential Quadratic Programming (SQP)([13]) is used in every iteration of the goal programming to solve the non-linear equations. The problem under consideration involves only equality constraints. Every SQP problem has a quadratic programming sub-problem. The objective function 'F' and the equality constraint 'h' are converted into a Lagrangian as

$$L = F + \lambda h \tag{6.16}$$

The divergence δL and the Hessian $\delta^2 L$ of the Lagrangian are calculated in each step. The matrix 'N' is calculated by determining the derivative of the constraints. The first order approximation is as follows

$$\nabla L(x, \lambda) = \nabla L(x_0, \lambda) + \nabla^2 L(x_0, \lambda) \begin{pmatrix} \delta x \\ \delta \lambda \end{pmatrix} \tag{6.17}$$

$$\begin{pmatrix} \nabla^2 L(x_0, \lambda) & N \\ N^T & 0 \end{pmatrix} \begin{pmatrix} \delta x \\ \delta \lambda \end{pmatrix} = \begin{pmatrix} \nabla L(x_0, \lambda) \\ h \end{pmatrix} \tag{6.18}$$

The results of the optimization with the use of goal programming technique for the problem under consideration are shown in Figure 6.3.

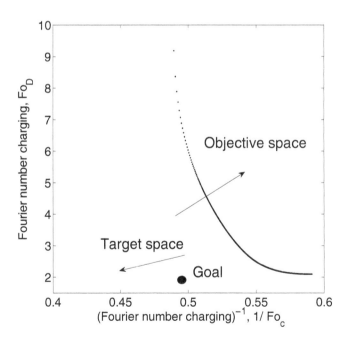

FIGURE 6.3
Pareto optimal points obtained from goal programming algorithm.

From Figure 6.3 it is evident that no solution in the target space lies in the objective function space. So the main objective of the goal programming can now be re-visited and stated as minimizing the deviation of the function value from the stated goal. Goal programming finds a single solution in each iteration for the given weights. This algorithm requires a priori information from the decision maker. The weighting vectors used here control the under attainment or over attainment of the goal. A total of 100 non-dominated solutions were generated. Each solution is a result of the change in the weight vector w_i.

6.7 Results obtained with non-dominated sorting genetic algorithm—NSGA-II

The elitist Non-Dominated sorted genetic algorithm (NSGA) also known as NSGA-II ([19]) is used to simultaneously to optimize both the objectives. Rudolph's elitist multi-objective evolutionary algorithm developed in 2001 implemented the elite preserving mechanism and the solutions converged to the true Pareto curve in a finite number of generations. In NSGA-II (Figure 6.4) not only are the elite members in each generation carried over to the next generation, but along with this the diversity of the solution is also preserved in each generation.

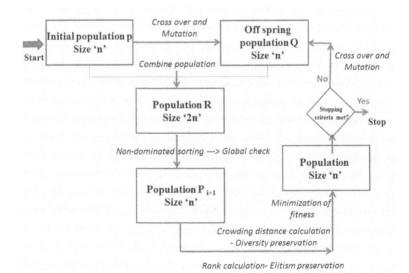

FIGURE 6.4
Flowchart for the NSGA-II algorithm employed in this study [65].

As a first step in the algorithm, the offspring population is generated from the parent population. The parent and the offspring are combined and a new population is formed. The key difference between GA and the NSGA-II is that in the former the parents compete among themselves. However, in NSGA-II the parent population and offspring population compete with each other to survive in the next generation. The best ranked solutions from the current generation are carried over to the next generation. This procedure is repeated until the Pareto optimal front is reached. In other words, the algorithm is executed until the members in the previous and current generation do not change.

The output of NSGA-II can be evaluated by two parameters:

- Closeness to the true Pareto solution (Obtained from brute force search which is explained in a later section)

- Diversity in the solution

Additionally, the effect of crossover and mutation on the diversity of the solution has been studied. The following method is used for crossover and mutation in NSGA-II

- Crossover - Simulated Binary crossover:

- Mutation - Polynomial mutation

$$c_{1,k} = \frac{1}{2}[(1 - \beta_k)P_{1,k} + (1 + \beta_k)P_{2,k}] \tag{6.19}$$

$$c_{2,k} = \frac{1}{2}[(1 + \beta_k)P_{1,k} + (1 - \beta_k)P_{2,k}] \tag{6.20}$$

$c_{i,k}$ is the ith child with k^{th} component. β_k is a random number; $p_{i,k}$ is the selected point.

$$p(\beta) = \frac{1}{2}(\eta_c + 1)\beta^{\eta_c} \tag{6.21}$$

$p(\beta)$ is the probability distribution of β. η_c is the distribution index of crossover that will determine how well spread the children will be from parents. The cross over fraction P_c is defined as the fraction of individuals in the next generation, other than the elite children that are created by crossover. The mutation fraction p_m is defined as the fraction of individuals in the next generation, other than the elite children that are created by mutation. In this study,

$$P_c + p_m = 1 \tag{6.22}$$

The effect of crossover is clearly seen in Figure 6.5. Initially, as the rate of crossover is increased the average Pareto distance also increases. However, after $p_c = 0.6$ an increase in crossover leads to a decrease in the average Pareto

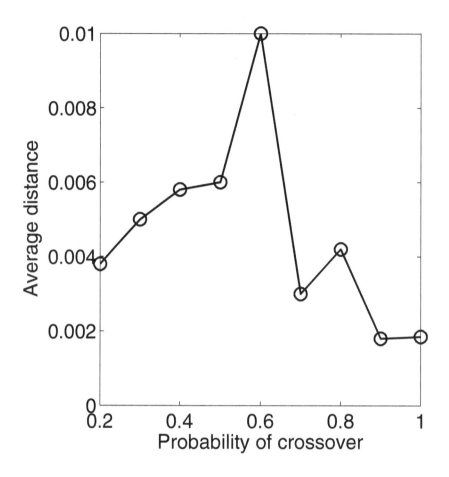

FIGURE 6.5
Effect of probability of crossover on the diversity of the solutions obtained
using NSGA-II algorithm [65].

distance. The average Pareto distance is the measure of the distance between
two individuals in the solution. This again is an indication of the diversity
of the solution. For any random number of generations, the diversity of the
solution is high at $p_c = 0.6$. At higher cross over rates, a lot of intermediate
points are generated between the solutions at each generation, thus leading
to a lower diversity in the solutions obtained. After the crossover probability
is fixed, the number of generations required is to be decided.

For G=500, the maximum diversity beyond this did not increase. From
the sensitivity analysis it can be concluded that a total of 500 generations
and $p_c = 0.6$ are required to obtain better solution in terms of diversity and
closeness to the true Pareto curve.

6.8   Particle swarm optimization

Multi-objective particle swarm optimization (MOPSO) was developed first in 1995, inspired by an animal social behaviour simulation system that incorporated concepts such as nearest-neighbour velocity matching and acceleration by distance (Kennedy & Eberhart, 1995). MOPSO utilizes a population called a swarm. The swarm moves through the objective space, searching for 'better' solutions, with its position modified stochastically with each iteration. This modification of the swarm differs significantly from evolutionary algorithms since it reflects a cooperative nature instead of a competitive nature. In order to modify the population and favor the better performing individuals, MOPSO updates the velocity vector for each member of the swarm, using this to shift its position at each iteration of the algorithm. The update in velocity and thereby position has both an elitist component as well as a stochastic component. The elitist component uses information about the best position encountered by the particle in its previous experience, as well as information about the best position encountered in the entire swarm's previous experience, to update the velocity. The stochastic element consists of the weight allotted to each of these information sources about previous good positions before it is used to update the velocity. The particles hence move towards promising regions of the objective space by using information derived from their own experience as well as the experience of other particles.

6.9   Brute-Force search

After performing multi-objective optimization, to get an idea of the objective function space and benchmark the results obtained, the fitness function generated through ANN is evaluated at 10^5 design points using Matlab R14. Latin hypercube sampling technique was used to generate the random points. In this study, the brute force optimum is used as a proxy for the true optimum, though this too is accurate only to the extent ANN is.

For a minimum-minimum problem the left border of the objective function space Figure 6.6 corresponds to the true Pareto curve.

Thus, the exhaustive search method gives an idea about the feasible objective space as shown in Figure 6.6 and can be used as a benchmark tool to measure the performance of other multi-objective optimization algorithms for a given problem. The computational cost and accuracy in this search method are directly proportional to the number of candidate solutions. Now both NSGA-II and goal programming solutions are taken and compared with the true Pareto curve obtained from the brute-force search. For NSGA-II, an increase in the number of generations beyond 500 did not result in a change in closeness of

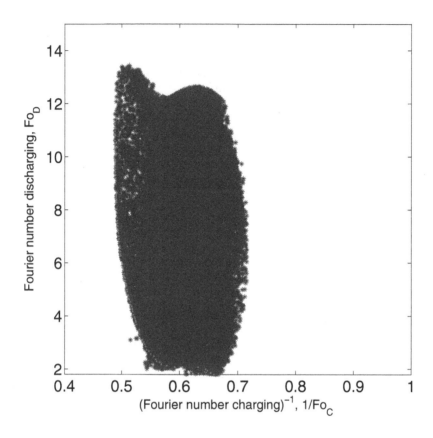

FIGURE 6.6
Objective function space obtained with the brute force search method.

the solution to the true Pareto optimal solutions. As far as goal programming is concerned, every solution lies on the true Pareto curve as the objective of the algorithm is to find the solution which is closest in the objective function space.

6.10   Clustering of Pareto solutions

In general, it is difficult to identify a particular solution from the set of design solutions in a multi-objective optimization problem. The main goal of clustering algorithms is to help the designers identify groups in the solution set that correspond to various levels of goal attainment with respect to each objective. The clusters help the designer identify the areas where distinct solutions achieve the same level of objectives. Hence two clustering approaches namely K-means clustering and Technique for Order of Preference by Similarity to Ideal Solution (TOPSIS) are combined together to achieve this goal. This process is tedious, but this technique helps the designer identify the level of initial and end point of each cluster with respect to two objectives. This technique will aid the decision maker to understand a large solution set rather than a large number of individual solutions.

k-means clustering is a well known simple non-supervised clustering algorithm widely applied in the solutions of multi-objective optimization algorithms. 'k' denotes the number of clusters, and hence in this algorithm, initially k centers are randomly defined each for one cluster. The centers are placed as far as possible from each other. In the following step, each point is associated with the nearest center. The algorithm aims to minimize the observation data to the centroid. In the next iteration, the mean or the centre point of each cluster is assigned as centroid and this procedure is iterated until the centroid does not change with iterations.

Technique for Order of Reference by Similarity to an Ideal Solution (TOPSIS) [84] aims to minimize the distance from the ideal solution and maximize the distance from the negative ideal solution. In this study, the positive ideal solution is [0.4881 1.832] and the negative ideal solution is [0.65 11.26].

The distance L_p is calculated by using Minkowski's method

$$L_p = \sum_{i=1}^{k} w_i^p |f_i - f^*|^p \tag{6.23}$$

where f* is a vector which corresponds to the ideal solution. The value of p varies between 1 and ∞. w_i is the weighting vector.

The steps involved in the TOPSIS algorithm are as follows:

- Create a matrix a_{ij} for the set of non-dominated solutions obtained from NSGA - II.

- Normalize the matrix

$$r_{ij} = \frac{a_{ij}}{\sqrt{\sum_{j=1}^{k} a_{ij}^2}} \tag{6.24}$$

- Calculate the weighted normalized values for each element in the normalized value

$$v_{ij} = w_i r_{ij} \tag{6.25}$$

where w_i is the weight set by the decision maker.

- Determine the positive ideal solution (I_p) and the negative ideal solution (I_n).

Maximize the distance from the negative ideal solution and minimize the distance from the positive ideal solution.

- Calculate the Euclidian distance of each element of each v_{ij} from the positive ideal solution.

$$d_j^+ = \sqrt{\sum_{i=1}^{n}(v_{ij} - I_p)^2} \tag{6.26}$$

- Similarly calculate the distance from negative ideal solution(d_j^-).

- Calculate the closeness of each solution of matrix 'v' to the positive ideal solution by

$$D_j^+ = \frac{d_j^-}{d_j^+ + d_j^-} \tag{6.27}$$

- The change in weights helps us determine the more appropriate solution from a set of non-dominated solutions.

The TOPSIS algorithm [84] was coded in Matlab and the result of the algorithm is shown in Figure 6.7.

The clustering of the obtained Pareto optimal solutions is shown in Figure 6.7. The solutions are divided into 5 clusters. The clusters thus form the higher order information for the designer to choose one solution from the set of solutions. Cluster A corresponds to the highest charging time and discharging time. However, cluster E corresponds to the lowest charging and discharging time. Figure 6.8 shows the diversity of the final solutions based on the 3 algorithms. The brute force has an equal fraction of solutions in all of the 5 cases.

Furthermore, it is evident that the other 3 algorithms have a distinct spread of fractions over the clusters. Goal programming has the highest diversity of fraction in between the clusters. This is in fact not a good measure for a multi-objective optimization problem. After a sensitivity analysis it is very clear that NSGA-II solutions have a good diversity which is clearly depicted as an almost equal spread of the fraction of solutions among the clusters.

6.11 Discussion

From the carefully performed optimization procedure elucidated above, it is evident that the NSGA-II algorithm converges very close to the global Pareto

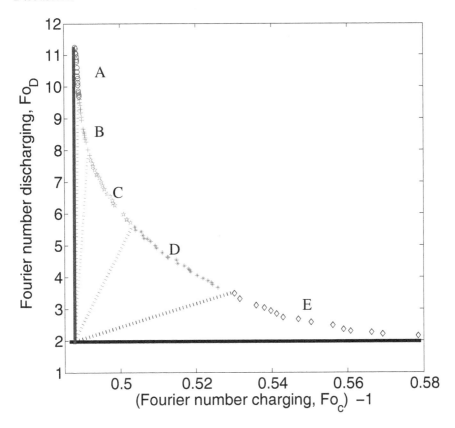

FIGURE 6.7
Clustering of Pareto optimal solutions for the problem of discrete heating of
72 pin fin heat sink [65].

optimal solution with a very minimum number of initial points. However, a
sensitivity analysis is required to arrive at the optimum number of generations
and the probability of crossover and mutation. This algorithm is more stable
to any change in the initial population.

Discussion of results can be done for two different cases as shown in Figure 6.9.

- only melting

- melting and solidification

We should reiterate the fact that the basic assumption in this study is that
we, the thermal engineers, have the freedom to decide the hotspot distribution
and explore which hotspot configuration is better.

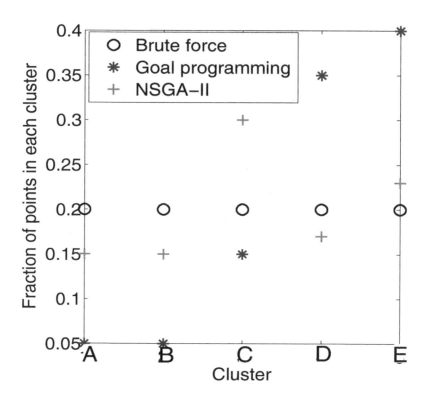

FIGURE 6.8
Diversity measure of optimization algorithms.

Based on this if only melting is considered, it is evident that if any of the two heaters receive 10% of the total power, the melting is enhanced and is best next only to the uniform heating situation. However, when solidification is also considered, the optimal configuration of discrete heaters is not the same as the "only melting" case. Even if the uniform heating case is included in the optimization, the discharge time for this is very high and is not the best solution when both the objectives are considered. This is a surprise result. From the multiple solutions (Pareto optimal solutions), five representative solutions are chosen using the TOPSIS algorithm and their thermal performances are analyzed. From the initial population of non-uniform discrete heating, the best possible melting time is determined to be 5255s. The corresponding discharge time is 17860s. Among the five representative solutions, the solution with least melting time of 5534s outperforms the best candidate from the initial population. The time to melt is thus increased by 10%. Furthermore, the corresponding discharging cycle time of the optimal solution is 13718s, which is 23.5% smaller than the best melting case in the initial population.

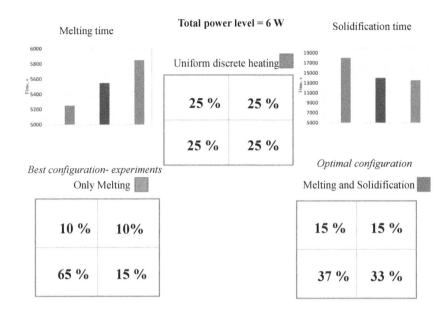

FIGURE 6.9
A bird's eye view of the optimal configurations with conflicting objectives [65].

On close observation, it is seen that for this improvement in thermal performance enhancement, two among the heaters are receiving 15.3% and 14.6% of the total heat. This can be closely approximated as two heaters receiving 15% of the total heat load. The other two heaters receive 37 and 33% respectively in a single chip module.

From this study, it is evident that the brute-force search method gives a picture of the objective function space and hence a clear idea of the true Pareto optimal solution. The major drawback of the brute force search method is the large number of points required to generate the objective function space and high computational time. However, this method can be used to test any multi-objective algorithm, in terms of its closeness to the true Pareto optimal solution. Table 6.4 reports a critical comparison of the performance of the four optimization algorithms implemented in this study.

The NSGA-II algorithm is seen to converge very closely to the global Pareto optimal solution with a very minimum number of initial points. The computational time is the second least among the three algorithms discussed above. However, a sensitivity analysis is required to arrive at the optimum number of generations and the probability of crossover and mutation. This algorithm is more stable to any change in the initial population.

TABLE 6.4

Comparison of the performance of the three optimization algorithms used in the present study

SI No	Parameter	Brute-force	NSGA-II	GP	PSO
1	Initial points	10^5	48	1	48
2	CPU time, s	77	11.55	65	9.88
3	Closeness to true solution	Benchmark	yes	yes	yes
4	Sensitive to	number of candidate solutions	Pc,Pm,G	initial guess	swarm population
5	Optimization method	Objective function based	Objective function based	Calculus based	Objective function based
6	Diversity of solution	Serves as benchmark	Preserved	Steps of the weight	Initial population
7	Hypervolume Indicator	0.6031	0.5932	0.5881	0.5992

Goal programming algorithm converges to the True Pareto optimal solution. The goal programming technique is highly time consuming and unstable to the initial guess values. Additionally, to obtain the value of the goals in the goal programming, a prior knowledge of the goal is required or it has to be derived by performing an additional single objective optimization. The NLP solution is highly sensitive to the chosen initial guess. The solutions are more or less equally spaced having uniform diversity throughout.

On comparing the limits of the favourable points obtained by MOPSO with those obtained by brute force and NSGA II, it is seen that MOPSO is not only able to find more favourable points, but it is also able to cover the entire range of possible favourable points much better. In comparison with NSGA-II, the ranges covered are quite similar.

The circuit board designer is now aware that the distribution of heat on the PCB is an important factor in thermal control by the PCM-based heat sink. He or she can attempt to modify the heat distribution to an appropriate one on the circuit board to improve the charging-discharging cycle characteristics of the PCM-based heat sink. The heat sink designer is also aware that predicting the performance of a heat sink independent of the distribution of heat sources on the PCB involves error. The heat sink designer also has evidence here that the assumption of a uniform heating source is not an optimum assumption and that the optimum designs are considerably far from the uniform heating case. Hence, it is more appropriate to model an approximation of the heat distribution at least for applications that are sensitive to the heat rejection and charging-discharging cycles.

6.12 Conclusions

Experiments were conducted on a phase change material (PCM)-based 72 pin heat sink with 4 identical discrete heaters at the base. The heat sink was made of aluminum and the PCM material used was n-eicosane.

Four different distinct algorithms were used to determine the set of Pareto optimal solutions for two conflicting objectives namely

$$Objective\ 1\ =Maximize\ charging\ cycle\ time$$

$$Objective\ 2\ =Minimize\ discharge\ cycle\ time$$

The important conclusions from the present study are

- The brute force search method can be used as a benchmark tool to evaluate the performance of other multi-objective optimization algorithms. Using a NSGA-II multi-objective optimization algorithm, a set of non-dominated solutions will be obtained instead of a single optimum solution.

- Goal programming was also implemented to obtain the non-dominated solution for the problem under consideration. It was found that the high non-linear nature of the problem makes the algorithm computationally very expensive. Additionally, the algorithm needs prior information about the ideal solution (Goal) and the convergence of the algorithm is highly sensitive to an initial guess.

- Among all the algorithms studied, NSGA-II is more reliable and has a better diversity.

The results from the present investigation on multi-objective optimization of the discrete power level of a pin fin heat sink with PCM can pave the way to better utilization of NSGA-II multi-objective optimization algorithms for highly non-linear problems. Henceforth, for the studies to follow, on the problems of this class the NSGA-II algorithm is recommended.

6.13   Closure

This chapter discussed the thermal optimization of the 72 pin fin heat sink using multiple candidate multi-objective algorithms. The optimal obtained was verified by conducting in house experiments. In the next chapter, results of investigations on a new breed of matrix pin fin type heat sink, an alternative to the pin fin heat sink, discussed in this chapter, will be presented.

7

MULTI-OBJECTIVE GEOMETRIC OPTIMIZATION OF A PCM-BASED MATRIX TYPE COMPOSITE HEAT SINK

7.1 Introduction

This chapter reports the results of experimental, numerical and geometric optimization of a matrix type heat sink, a potential alternative to the 72 pin fin heat sink considered in Chapter 5. A uniform heat flux is considered in this study. The baseline comparison cases in this study are the no fin or the pin fin case. Based on preliminary experiments, a matched up numerical model is first developed to better understand the melting and solidification of PCM inside the heat sink. This is a huge adavantage that a numerical model offers in that it lets one conduct virtual flow visualisation studies. The results discussed in this chapter are based on from the work reported in [69].

7.2   Experimental setup

A pin fin matrix type heat sink (comprising both vertical and horizontal fins) made of aluminum was used in the present study. The overall dimensions of the heat sink were 80 mm x 62 mm x 25 mm. The heat sink cavity has a depth of 20mm. The fin thickness is 2 mm in all the directions. A wall thickness of 7mm is maintained in the heat sink. A 2 mm slot is given at the bottom of heat sink to accommodate the plate heaters. All four sides of the heat sink are insulated by using a rubber cork and the top portion of the heat sink is covered with acrylic. The portion between the acrylic and heat sink top is sealed by using a silicon rubber gasket. The dimensions of the heater plate are 60 x 42 mm^2. The heater is made of mica sheet and nichrome wire of finite resistance is wound over it.

The pin fin matrix consists of 24 vertical fins, 6 full length horizontal fins in one perpendicular direction and 16 full length horizontal fins in the other direction, as shown in Figure 7.1. The dimensions of all the three configuration cuboidal heat sinks are given in Appendix A.

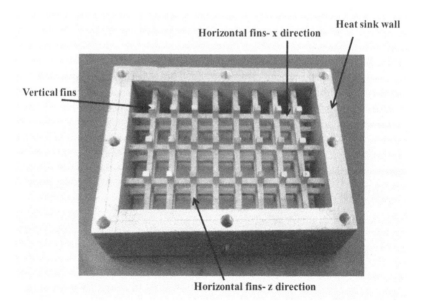

FIGURE 7.1
Photograph showing the heat sink before assembly [69].

The fins act as Thermal Conductivity Enhancers (TCE) and have a volume fraction of 8%. This volume fraction was chosen in order to have a direct one-to-one comparison with the of 72 pin fin heat sink. The pin fin matrix is fabricated by the process of Electro Discharge Machining. The side walls are welded to the matrix by gas welding. A leak test was performed initially on the heat sink, and when no leak was observed the experiments were started. The final assembled heat sink is shown in Figure 4.1. A total of 13 calibrated k-type thermocouples are used to measure various temperatures in the heat sink. Calibration is carried out using a constant temperature bath in the temperature range of 30 to 70°C. Two thermocouples (TB_1 and TB_2) are placed at the heat sink base. Four thermocouples (TW_1 TW_2 TW_3 TW_4), each are placed on each wall. Two thermocouples, each are placed at 5mm (T_{51},T_{52}), 10mm (T_{101} T_{102}), and 15mm (T_{151} T_{152}) from the base inside the PCM. One thermocouple (T_{amb}) is placed in still air to measure the ambient temperature. Figure 7.2 shows the position of thermocouples. The thermocouples are fixed to the heat sink using Araldite^{TM} epoxy. All the thirteen thermocouples are connected to an AGILENT 34970A data logger.

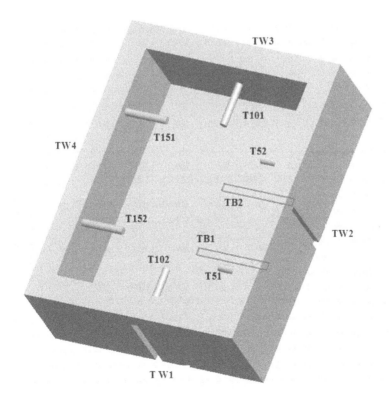

FIGURE 7.2
Picture showing the position of thermocouples in the heat sink [63].

The heating cycle is carried out until a set point temperature of $45°$ C is reached.

7.2.1   Uncertainty analysis

The uncertainty associated with the measurement of fundamental quantities like the voltage and current which are the least counts of the corresponding measurement devices. The temperatures are accurate to within $\pm 0.2 ° $ C.

The uncertainty in the power can be calculated as

$$\sigma_p = \pm\sqrt{\left(\frac{\partial P}{\partial V}\sigma_V\right)^2 + \left(\frac{\partial P}{\partial I}\sigma_I\right)^2} \tag{7.1}$$

for a power level of 2W, the uncertainty in the power turns out to be

$$\sigma_P = \pm\sqrt{(0.5 \times 0.1)^2 + ((4 \times 0.01)^2} \tag{7.2}$$

$$\sigma_p = 0.064 \tag{7.3}$$

$$\frac{\sigma_P}{P} = 0.032 \tag{7.4}$$

$$\sigma_P = 3.2\% \tag{7.5}$$

7.3   Charging and discharging cycles

As discussed earlier, the operation of a heat sink takes place in two cycles, namely charging and discharging cycles. Charging here refers to the process of energizing the heat sink base heater plate with a constant DC power supply. In the present study, charging is done until the heat sink base reaches a set point temperature of $45°$C. Discharging here refers to switching off the DC power input to the heat sink base. During this process, the heat sink loses heat to the ambient air through natural convection. Experiments are conducted for different power levels ranging from 3 to 10 W as shown in Figure 7.3. As the power input increases, the time taken by the heat sink to reach a particular set point temperature decreases.

As seen from the Figure 7.3, for a power input of 4W the time taken to reach the set point temperature of 318.2K is 5620s, while for a power level of 6W the time taken is 2880s. For a power input of 10W, the time taken is considerably reduced to 1450s which is one-fourth of the time taken for the 4W case.

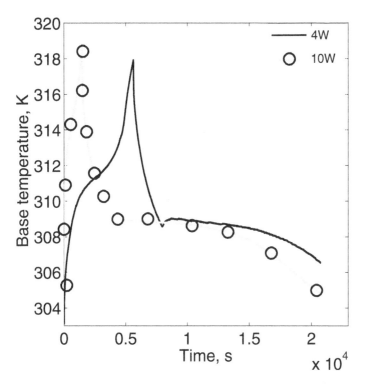

FIGURE 7.3
Variation of the base temperature with time for 7 different power levels for
the PCM-based matrix type heat sink [69].

7.4 Baseline comparison of heat sink with PCM to that of heat sink without PCM

Comparisons are done in the present study, for the heat sink with PCM against
a heat sink without it, as shown in Figure 7.4. For the purpose of comparison
a new term 'ER' is defined (enhancement ratio) as the ratio of the time taken
by the heat sink to reach a set point temperature with PCM to that without
PCM. The enhancement ratio 'ER' for a power level of 5W is 9.80. As the
power level is increased the enhancement ratio decreases.

Taking the experimental temperature measurements as benchmark, a
numerical model is developed using the same procedure as discussed in Chapter 5. The temperatures at various locations in the heat sink are matched
with the experimental measurements and an overall heat transfer coefficient
is calculated.

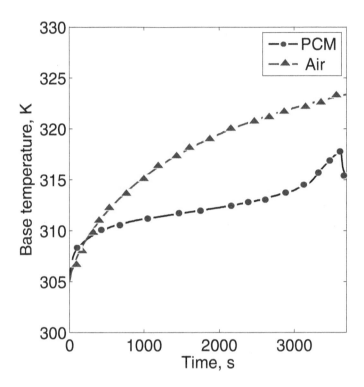

FIGURE 7.4
Comparison of the experimentally obtained temperature-time history for the
heat sink with PCM for a power level of 6W with that of heat sink with air
[69].

For a power level of 6W, the numerical simulations yielded some interesting
results as shown in Figure 7.5.

When the heater temperature is compared, the no fin case showed a sudden steep increase after t=200s. Between, the pin fin case and matrix type
similar trends are observed. On observing the PCM volume temperature till
1000s of heating, the PCM volume temperature is comparable for all 3 cases
up to t=1000s. Beyond 1000s of heating, there is a large deviation in temperatures for the no fin case with respect to the other two cases. Temperature
contours for the no fin case are shown in Figure 7.6. From this figure, it can be
inferred that the gradients are very steep in the vertical direction for the no fin
case.

A heat sink without a fin becomes a direct victim of the "self-insulation"
effect. A temperature difference of 35 K in 20mm height results in a temperature gradient of 1750 K/m. Furthermore, across the radial direction, the heat
propogation is a lot poorer.

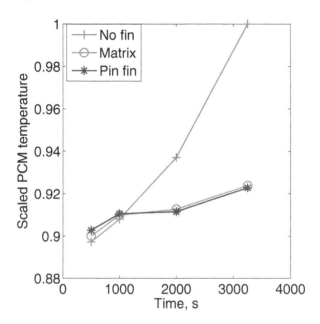

FIGURE 7.5
Comparison of time history of scaled PCM temperature for the no fin, 72 pin fin and matrix fin case.

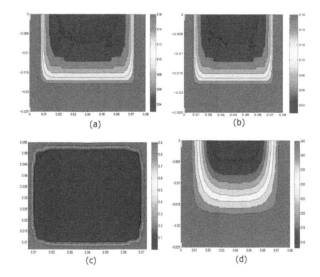

FIGURE 7.6
Contours of (a) temperature after 200s of heating, (b) temperature after 1000s of heating, (c) liquid fraction after 2000s of heating, (d) temperature after 3250s of heating (all for a no fin case).

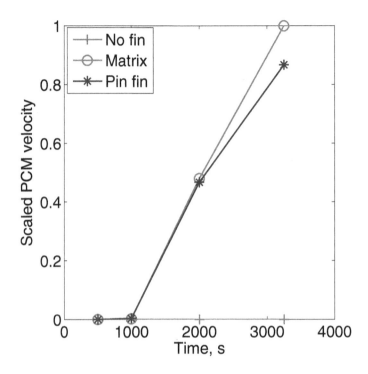

FIGURE 7.7
Comparison of transient history of scaled PCM velocity for the no fin, 72 pin
fin and matrix fin case.

For the no fin case, the only mode of heat transfer is conduction. Figure
7.7 shows the ratio of PCM velocity to the maximum velocity encountered in
the PCM volume. It is seen that the no fin case shows zero velocity through
the melting process. The evolution of velocities for pin fin and matrix fin heat
sinks follows similar trends up to t=2000s. The matrix heat sink shows a 15%
higher velocity at t=3000s. For a matrix type heat sink, even at early stages of
melting, one can find that a small mass of top layer PCM has already melted
(See Figure 7.8). This is a good indication that the thermal gradient prevailing
in both the directions is not very high.

The heat from the base and the side walls diffuses out equally into the
PCM. However, in the 72 pin fin case, as shown in Figure 7.9 there is not
much diffusion of heat in the radial direction. After 2000s of heating, there is
a slight increase of scaled PCM temperature for a matrix fin case.

This is a strong indication that the matrix fin case is now a victim of
quick melting. This result again reiterates the fact that even though the same
amount of TCE exists in both the 72 pin fin and the matrix case and the
distribution of TCEs affects the melting process. This is the much addressed
trade-off in the literature, between quick melting (Figure 7.10) and effective

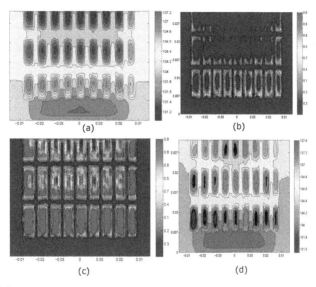

FIGURE 7.8
Contours of (a) temperature after 200s of heating, (b) liquid fraction after 1000s of heating, (c) liquid fraction after 2000s of heating, (d) temperature after 3250s of heating (all for a matrix fin case).

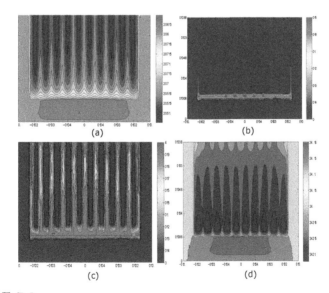

FIGURE 7.9
Contours of (a) temperature after 200s of heating, (b) liquid fraction after 1000s of heating, (c) liquid fraction after 2000s of heating, (d) temperature after 3250s of heating (all for a 72 pin fin case).

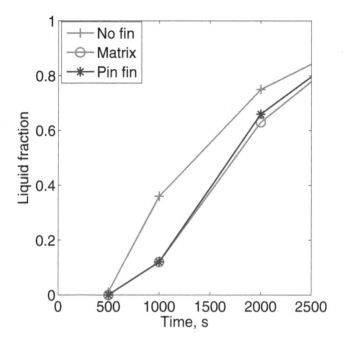

FIGURE 7.10
Comparison of transient history of liquid fraction for the no fin, 72 pin fin and matrix fin case.

utilisation of latent heat. From the standpoint of electronic cooling, quick melting is disadvantageous as the heater temperature quickly rises (sensible heating) post melting.

However, an exactly opposite behaviour is observed in the solidification. In an electronic cooling application the melting and solidification objectives are often conflicting. The matrix type fin heat sink cools faster than a 72 pin fin heat sink. This reiterates the fact that the only mode of heat transfer during solidification is conduction. The presence of TCEs in both directions helps the matrix heat sink to solidify faster. As a logical extension to this finding, geometric optimization of the matrix heat sink is now carried out. The NSGA-II algorithm was established as the best among the four considered in Chapter 6. Hence for this study, multi-objective optimization has been carried out with NSGA-II.

7.5   Numerical model

Using numerical simulations to conduct an extensive parametric study is critical to determine the optimal configuration of the heat sink for maximum performance, in view of the fact that detailed parametric studies through experimentation are both time consuming and expensive. The modeling of phase-change processes presents a significant challenge due to the complexity and conjugate nature of the phenomenon involved ([24]). Some important factors that need to be considered in the modelling are

- volume expansion during phase change

- convection in liquid phase

- motion of solid in the melt due to density differences

In the current study, the functioning of the heat sink with phase change material is simulated through the ANSYS FLUENT 14.0 ([75]) and the results of validation for the base temperature time history subject to a power level of 6W at the base are shown in Figure 7.11

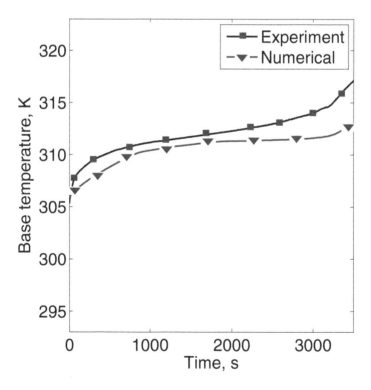

FIGURE 7.11
Validation of the numerical model with experiments for matrix pin fin case [69].

7.5.1   Geometry and mesh

Figure 7.12 shows the mesh used in obtaining the numerical solutions. The dimensions are the same as those used in the experiments. The model is meshed using tetrahedral meshing on aluminum material and hexahedral mesh on PCM. The mesh element sizes are 1 and 2 mm on the PCM and aluminum sides respectively. The maximum face size and the maximum size of the grid are 1.2 and 2.4 mm respectively. The orthogonal quality of the meshing is 0.87 and is shown in Figure 7.12. The numerical model and methodology are same as the ones discussed in Chapter 5.

7.5.2   Grid independence studies

The heat sink is meshed with tetrahedral meshes on the aluminum material and with hexahedral mesh on the PCM. To determine the number of elements that need to be taken into account for simulation, a grid independence study was conducted. It is observed that a mesh with 284,000 nodes is adequate and this was used for all subsequent computations.

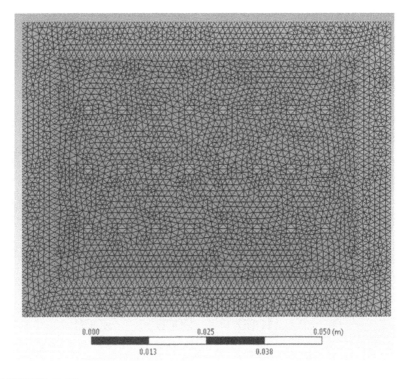

FIGURE 7.12
Figure showing the mesh employed for the matrix type heat sink [69].

7.5.3 Boundary conditions

The heat sink is subjected to two cycles viz. the heating and the cooling cycles. Heat losses occur around the heat sink owing to natural convection and these need to be quantified. Although the heat transfer coefficient varies with the surface temperature, an overall average heat transfer coefficient as shown in Figure 7.13 is determined by comparing the temperature time histories simulated for various values of U with respect to the experimental temperature time history.

For this, the heating and the cooling cycles are treated separately and the overall heat transfer coefficient is varied in steps of $0.5W/m^2K$. Figure 7.14 shows the variation of sum of the square of the residuals R^2 with the heat transfer coefficient for the heating cycle and a similar plot obtained for the cooling cycle.

However, to develop the numerical model, five thermocouple measurements which include the temperature reading inside the PCM were also considered to

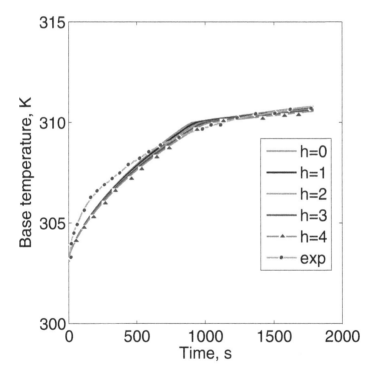

FIGURE 7.13
Comparison of transient history of base temperature for various assumed values of heat transfer coefficient with experiments for a power level of 6W at the base.

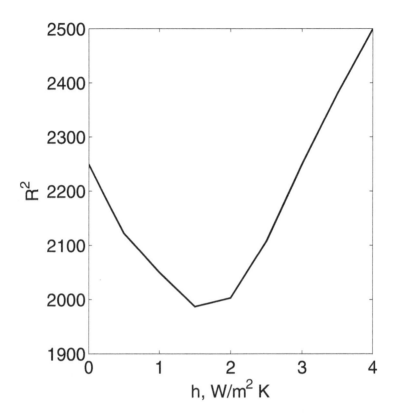

FIGURE 7.14
Variation of sum of residuals, (S) with the heat transfer coefficient(h) for the heating cycle [69].

estimate the overall heat transfer coefficient (U). The value of U is determined by minimizing the sum of the square of the error in the temperatures in a least square sense. Figure 7.14 shows the variation of the sum of the residuals, S against U and it is seen that U=1.5 W/m^2K has the least error.

Initial conditions

At t=0 (was set in accordance to the experiment)

$$T = T_{Al} = T_F = T_S = 305.5K \qquad (7.6)$$

Boundary conditions

For the top surface of the heat sink

$$\frac{\partial T}{\partial y} = \frac{\partial T_{Al}}{\partial y} = \frac{\partial T_F}{\partial y} = \frac{\partial T_S}{\partial y} = 0 \qquad (7.7)$$

for all the other outer surfaces

$$\frac{\partial T_{Al}}{\partial y} = 0 \tag{7.8}$$

For the surface in contact with the heater

$$-k_{Al}\frac{\partial T_{Al}}{\partial y} = q \tag{7.9}$$

7.6 Optimization

The goal of this investigation is to determine the optimal configuration of the heat sink with two objectives viz. maximizing the heating time and minimizing the cooling time, given the constraint that the volume of the phase change material is constant.

The following steps are used to carry out the optimization:

1. Generation of geometries with random dimensions using Latin hypercube sampling (Discussed in detail in Chapter 6)

2. Performing numerical simulations for each geometry

3. Training the neural network to determine the input-output correlation

4. Using a multi-objective optimization tool to obtaining the optimal outcomes in the form of a Pareto plot.

It is instructive to note the symmetry of the problem in terms of the length and width. This, in effect, gives us 40 inputs to train the neural network. As the length and height are varied the spacing between the fins adjust accordingly keeping the percentage of TCE constant at 9%. The dimensions of the geometries thus generated are shown in Table 7.1.

For each of these geometries, simulations are done for the heating and cooling cycles to study the fluid flow and heat transfer behavior.

The variation of base temperature for the select four models is shown in Table.7.2. From the table, it can be observed that the case with highest charging time did not give the best discharging time. Furthermore, the case with the best discharging time did not give the best charging time. From the 40 models that were generated and monitoring their corresponding outputs, there seems to be a considerable diversity spread in both the charging and the discharging cycles. Hence, there appears to be an excellent scope for carrying out multi-objective optimization.

TABLE 7.1
Geometrical models generated using the LHS method [69]

ModelNo.	Length mm	Width mm	Height mm
1	62.9	32.0	31.5
2	65.3	47.2	20.6
3	60.5	37.6	27.9
4	59.3	33.6	31.8
5	73.1	50.4	17.2
6	61.1	39.2	26.5
7	67.1	39.2	30.3
8	64.1	44.0	22.5
9	55.1	44.8	22.1
10	58.7	48.8	22.1
11	56.9	48.0	23.2
12	53.9	43.2	27.2
13	66.5	52.0	18.3
14	56.3	49.6	22.7
15	64.7	52.8	18.5
16	71.3	42.4	21.0
17	68.3	34.4	27.0
18	52.1	45.6	26.7
19	70.1	40.8	22.2
20	71.9	53.6	16.4

7.6.1   Artificial neural network

The procedure for developing ANN was discussed in detail in Chapter 6.

The inputs to the ANN for the geometric optimization are the length (l) and height (h) of the PCM and the outputs are the heating time (T_h) and the cooling time (T_c) represented non-dimensionally as the Fourier number (F_o)

TABLE 7.2
Diversity in thermal performance for various geometries [69]

L(mm)	B(mm)	H(mm)	t_m (s)	t_s (s)
62.9	32.0	31.5	3960	18000[++]
52.1	45.6	26.7	4500[++]	24000
56.9	48.0	23.2	3625	20000
71.3	42.4	21.0	3295	21840

which are functions of the geometry of the heat sink viz. length (l), height (h) power level (Q) and the total volume of the PCM. The Fourier number is defined as

$$Fo = \alpha_{PCM} \frac{t}{H^2} \tag{7.10}$$

$$Fo = f(l, b, Q, V_t) \tag{7.11}$$

The optimization is carried out for the PCM volume (V_t) to be $63360mm^3$ for a constant power supply of 5W and a TCE volume fraction of 9%. Further, the bounds for the dimensions of the PCM are limited as follows

$$30 \leq l \leq 80mm \tag{7.12}$$

$$15 \leq h \leq 30mm \tag{7.13}$$

The input variables to the ANN are non-dimensionalized as l^* and h^* which are defined as

$$l^* = \frac{l}{l_{max}} \tag{7.14}$$

$$h^* = \frac{h}{h_{max}} \tag{7.15}$$

where $l_{max} = 80mm$ and $h_{max} = 30mm$. Thus, the non-dimensionalized values are in the following range

$$0.375 \leq l^* \leq 1 \tag{7.16}$$

$$0.5 \leq h^* \leq 1 \tag{7.17}$$

A neuron independence study was conducted to determine the optimum number of neurons to be used in the hidden layer. Performance parameters such as the root mean square error (RMS) and mean relative error (MRE) defined below, were calculated for networks constructed with varying nodes in the hidden layer. The parameters are defined as

$$MRE = \frac{\sum_{i=1}^{N} |t_{ANN,i} - t_{num,i}|}{N} \tag{7.18}$$

$$RMS = \sqrt{\frac{\sum_{i=1}^{N} (t_{ANN,i} - t_{num,i})^2}{N}} \tag{7.19}$$

Table 7.3 shows the variation of MRE and RMS with number of neurons in the hidden layer. From this study, it is seen that a total of 9 neurons in the hidden layer is adequate.

A two layer feed forward back propagation network is used and the network is trained using the Levenberg Marquadart algorithm in MATLAB.

TABLE 7.3

Results of the neuron independence study for the ANN employed in this study [69]

$SINo$	$Number of neurons$	MRE s	RMS s
1	1	8.15E-03	2.45E-02
2	2	3.81E-03	1.12E-02
3	3	4.08E-04	1.39E-02
4	4	5.41E-03	2.06E-02
5	5	5.98E-03	1.46E-02
6	6	2.08E-03	6.62E-02
7	7	7.22E-03	2.72E-02
8	8	1.61E-03	2.84E-02
9	9	8.45E-04	3.71E-03
10	10	6.69E-03	2.12E-02
11	11	4.07E-03	2.50E-02
12	12	8.86E-03	7.21E-02
13	13	3.92E-02	7.57E-02
14	14	1.54E-02	1.18E-02
15	15	4.81E-03	5.19E-02
16	16	1.45E-03	1.20E-02
17	17	5.13E-03	1.24E-02
18	18	5.85E-04	8.66E-02
19	19	2.72E-03	4.15E-02
20	20	2.32E-04	2.30E-02
21	21	1.53E-02	1.14E-02
22	22	2.88E-03	2.13E-02
23	23	7.44E-03	1.78E-02
24	24	2.28E-03	1.32E-02
25	25	1.53E-02	9.31E-02

The Levenberg Marquadart algorithm is used for a least squares curve fitting problem. For a given set of m empirical data points, input ($\mathbf{x}$) and output ($\mathbf{y}$) variables, the optimized value β is derived such that the sum of squares of the deviations ($S(\beta)$) is minimized.

$$S(\beta) = \sum_{i=1}^{m} [y_i - f(x_i, \beta)] \qquad (7.20)$$

Of the 40 data points, 80% data points are used to train the network and the remaining data points are used to validate the network. The R^2 during training was found to be 0.96, while that during the testing is 0.91.

7.6.2 Multi-objective optimization

In the present study two objectives, namely maximizing the heating time and minimizing the cooling time are considered.

$$Objective \ function \ f_1 = maximize \ charging \ time$$

$$Objective \ function \ f_2 = minimize \ discharging \ time$$

Subject to constraints

$$Cavity \ volume = 63360 \ mm^3$$

$$TCE \ volume - 6336 \ mm^3$$

In the current study, objective 1 is the time required for heating and objective 2 is the time required for cooling. The goal is to maximize objective 1 and minimize objective 2 simultaneously. MATLAB 2013b is used to implement the NSGA for multi-objective optimization. Since this solves for a minimization problem, the negative of objective 1 is used. Over 400 data points were generated to determine the Pareto front. Figure 7.15 shows the Pareto front obtained for the multi-objective problem. The figure shows a set of non-dominated solutions, each of which satisfies both the objectives. Thirty-two distinct points were obtained on the Pareto front. Four random Pareto optimal points are chosen from each of the clusters A,B,C and D as shown in Figure 7.15. The clustering of the data set was performed using the well known "k-means clustering" algorithm. The algorithm breaks the data set into k different clusters. 'k' random observations are chosen from the data set and are assigned as the seeds initially. All other observations are combined with the seeds based on the proximity of those observations from the seeds. The algorithm aims to minimize the distance between the observation data point and the seed point. After the set of clusters is formed the center point or the mean point of each cluster is assigned as the seeds in the next iteration. The process is continued until the mean in the previous iteration and the current iteration do not show much variation.

It is very clear from the Figure 7.15 that it is impossible to move from one point to another without degrading any one of the objective functions. For instance, data points belonging to the top most cluster A in the front satisfy the objective function 1 more as compared to the objective 2. In contrast the data points belonging to cluster D satisfy the objective 2 more as compared to objective 1. If more weightage is given to objective 1 points higher on the Pareto front need to be chosen and for a higher weightage of objective 2 vice versa. The multi-objective optimization yields a set of non-dominated solutions which could simultaneously increase the charging time and decrease the discharging cycle time.

Another important implication of Figure 7.15 is that 80% of the points generated on the Pareto front corresponds to the geometries of high surface area compared to the initial geometries. Hence, the increase in surface area

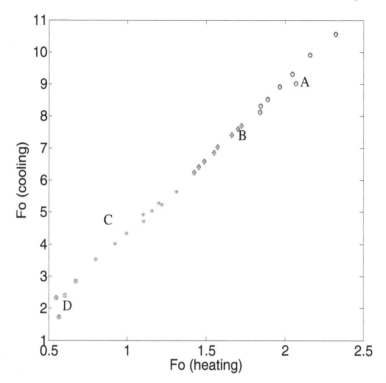

FIGURE 7.15
Pareto front obtained from the present study [69].

enhances the heat transfer during the discharging cycle and cooling time is minimized. Furthermore, the geometries corresponding to the points on the Pareto front have optimum spacing between the fins, which helps in full utilization of the latent heat of the PCM, thereby stretching the time of operation of the device.

7.6.3 Validation of optima

In order to have more faith in the optimization methodology proposed here, the numerically generated optima using the ANN-GA are validated by performing full numerical simulations. The optimized configurations predicted by the algorithm are verified using the validated numerical model that was developed using ANSYS FLUENT 14.0. The corresponding length, breadth and height are modeled and the estimated overall heat transfer coefficient was applied. The time taken to reach the set point temperature in heating turns out to be 3995 s(Fo=2.2), which is very close to the Fourier number of 2.4 predicted by hybrid ANN-GA. The results predicted by the optimization show a good agreement as shown in Figure 7.16

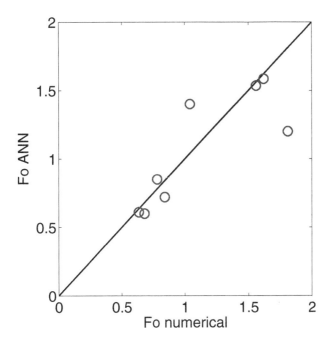

FIGURE 7.16
Parity plot for the time to reach set point obtained numerically with those obtained from the ANN [69].

7.6.4 Fluid flow and heat transfer characteristics of optimal configuration

The optimized case l= 51.46mm, b= 46.38mm, h=26.28mm (one among the non-dominated solutions favouring melting) for a heat input of 6W is analysed. The geometry corresponding to best charging time showed an enhancement of 12.5% during the charging time and 19% during the discharging cycle. Initially, heat conduction takes place from the heat sink body to the nearby PCM. At a particular time step isotherms are clearly placed near the aluminum surface. As observed from Figure 7.17, the layer of the PCM adjacent to the heater base gets heated up first and the temperature of that layer shoots up. Once the layer is fully molten, the adjacent layer is heated up and the heat flow progresses.

The propagation of heat is much faster in this matrix type heat sink due to the presence of both horizontal and vertical fins. From the liquid fraction contours, it is evident that the PCM near the heat sink base and some portions near the wall has completely melted at the end of 1000s . At the end of 2000s, the conduction of heat in both horizontal and vertical direction aids 50% of

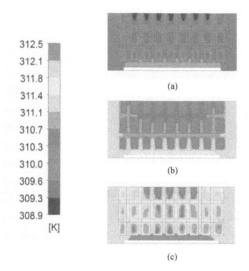

312.5
312.1
311.8
311.4
311.1
310.7
310.3
310.0
309.6
309.3
308.9
[K]

(a)

(b)

(c)

FIGURE 7.17
Temperature contours for optimized heat sink at time (a)500s (b)1000s
(c)2000s [69].

the PCM to melt. This is an excellent sign which indicates the utilization of
PCM latent heat to a larger extent.

At the end of 3600s, complete melting of PCM is observed. The disadvantage of having a matrix type heat sink can thus be inferred from the fact
that the entire PCM has melted at the end of 3600s, which limits the latent
heat time of the entire heat sink to 3600s. The merit of PCM-based heat sink
is largely based on the latent heat period, which for a matrix type heat sink
is limited due to the diffusion of heat in all the three directions. Figure 7.18
shows the solidification contours for the optimized configurations. One can
observe that the molten PCM close to the aluminum surface solidifies initially
forming an insulating layer for further heat transfer.

However, the operation time in this study is defined as the time taken
to reach a set point of 318.2K, which for a matrix type heat with optimized
configuration turns out to be more than 4100s.

In this study, the fin cross section is held constant, whereas the overall
length, breadth and height is varied. This indirectly is an optimization of the
spacing between the fins in the X, Y and Z direction. In Figure 7.19 L, B and
H are the overall dimensions. S_b, S_L are the spacing between the fins in the
X, Y, Z directions respectively.

The spacing between the fins plays an important role in the melting and
solidification phenomenon as the mode of heat transfer during melting is
both conduction and convection, while in solidification the only mode of heat

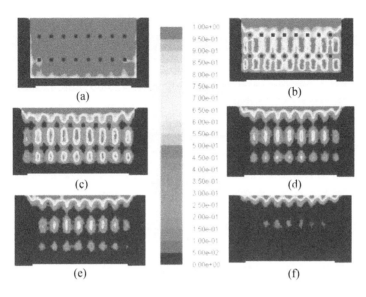

FIGURE 7.18
Liquid fraction contours for optimized heat sink during solidification at time
(a)10000s (b)135000s (c)175000s (d)21500s (e)24500s (f)25500s.

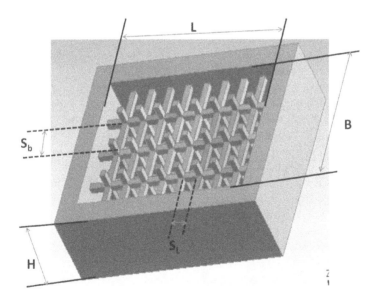

FIGURE 7.19
The key geometric parameters of the matrix pin fin heat sink [69].

transfer is conduction. If only melting is considered the spacing should be optimum enough to allow sufficient area for the convection currents, which originate from the heated surface (base, walls, fin surface) and which go on to erode the unmelted solid phase nearby and enhance melting of the PCM.

During solidification, it is better to have the aluminum surfaces close to each other so that more conduction takes place.

As soon as the heat flux at the bottom is made zero, the liquid PCM near the wall solidifies and forms a layer for solid PCM (conduction resistance). Due to the low thermal conductivity of PCM, the heat from the inside has to be conducted through this low thermal conductivity PCM layer and then reach the aluminum walls. As the time increases, resistance to the conduction is increased by the subsequent formation of solid PCM layers. In view of the above, it is clear that the solidification process needs less spacing between fins or more number of fins to break this insulating effect.

7.7   Conclusions

Experiments were conducted on a phase change material (PCM)-based matrix type heat sink heated from the base. A numerical model to mimic this was developed using commercially available ANSYS FLUENT 14.0 . Numerical results were matched up with experimental results for the case of uniform heat input of 5W to estimate the overall heat transfer coefficient which was then used in the optimization studies.

Subsequent to this, matched up full three dimensional numerical simulations of fluid flow and heat transfer were done for 40 different geometric configurations of the heat sinks ($30mm \leq l \leq 80mm$, $15mm \leq h \leq 30mm$)

The important conclusions from the present study are as follows

1. For a PCM-based matrix type heat sink, it is possible to use the NSGA multi-objective optimization approach to optimize the heat sink dimensions that can stretch the operation time of the heat sink during the charging cycle by 12.5% and reduce the discharging cycle time by 19% simultaneously.

2. Using NSGA multi-objective optimization algorithm, a set of non-dominated solutions was obtained instead a single optimum solution. The shape of the Pareto curve indicated the trade-off between the two objectives. Upon the analysis of the results, the physical characteristics of the phenomenon are well portrayed which help us understand the simultaneous influence of the input variables (l,b,h) on the performance of the heat sink. The cases corresponding to the optimal solution from each cluster were validated using the numerical model developed and excellent agreement was found.

3. This methodology adopted results in a set of non-dominated solutions. The algorithm can be extended to a wide class of problems, where optimization is required to satisfy two objectives which are equally important.

7.8 Closure

In the present chapter, the effect of variation of geometry on the matrix type heat sink was investigated. The performance of matrix type heat sink was compared with the no fin case and the pin fin case. A multi-objective geometric optimization of the matrix pin fin heat sink was carried out using the state of the art NSGA-II algorithm.

The results from the present investigation on multi-objective geometric optimization of a matrix type heat sink with PCM can

1. Serve as a benchmark for future numerical investigations

2. Pave the way for more focused research on optimization studies with simultaneous consideration of the charging and the discharging cycles.

The next chapter presents results of experimental investigations on cylindrical heat sinks, with an interesting variant, namely a cylindrical heat sink that rotates.

8

EXPERIMENTAL INVESTIGATION ON MELTING AND SOLIDIFICATION OF PHASE CHANGE MATERIAL-BASED CYLINDRICAL HEAT SINKS

8.1 Introduction

In Chapters 5 and 7, the results of investigations on rectangular composite PCM heat sinks were presented. In this chapter, results of experimental investigations of the effect of fins, gravity, rotational convection and mass of the phase change material (PCM) on the thermal performance of a heat sink subjected to constant heat flux of $5kW/m^2$ at the base are reported. This is not a pure passive heat sink case since the heat sink itself is rotating.

Three heat sink configurations with two media, namely air and n-eicosane are investigated to better understand the role of thermal conductivity enhancers (TCEs) on the melting and solidification heat transfer. The heat sink and fins are made of aluminum. Temperature measurements are carried out using calibrated wireless temperature transmission. A LiPo battery along with a potentiometer circuit is used to regulate the power input to the heater. The heat sink is subject to four fill ratios (0, 0.33, 0.66, 0.99) of PCM/air, nine orientations (0°,45°,90°,135°,180°,225°,270°, 315°, 360°), and three rotational speeds (0,60,120 RPM) simultaneously. The results from this book are a part of the patent filed by the author; (Indian Application number 201741018559) and [62].

8.2   Experimental setup

The photographic views of the heat sinks investigated in the present study are shown in Figure 8.1. Three heat sink configurations are investigated in the present study, namely (i) Heat sink with no stem and fins (i) Heat sink with a central stem (iii) Heat sink with internal stem and radial fins. The heat sink (iii) is fabricated using an electro discharge machining process. The heater is

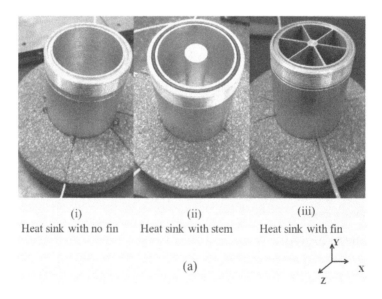

| (i) | (ii) | (iii) |
| Heat sink with no fin | Heat sink with stem | Heat sink with fin |

(a)

FIGURE 8.1
Photograph showing the three cylindrical heat sink configurations considered in this study [62].

made of mica sheet over which a nichrome wire is wound. The heater for all the three heat sinks is designed by considering the resistivity and the cross section of nichrome wire. The dimensions of all the three heat sinks are provided in Appendix A.

The diameter of the nichrome wire is 0.2 mm. Nichrome is chosen as it is highly corrosion resistant and does not oxidize at higher temperatures. The resistivity of the nichrome wire used in the present study is 1.5×10^{-6} Ω-m. Nichrome wire is wound over the mica sheet and is sandwiched between two mica sheets. The heater is initially designed to provide a resistance of 10 ohms. The length of the heater is determined to be 185 mm and it provides a maximum heat input of 10W at 10 V. The heater is attached to the heat sink surface by using a highly conductive heat sink paste. In order to avoid heat loss from the heater and also to promote uni-directional heat transfer into the heat sink, the bottom side of the heater is insulated by means of cork. The heat generated by the heating element depends on the current passing through the wire which again depends on the resistance, length, and cross section of the wire. The measurement of temperature is carried out using a wireless temperature measurement module and is calibrated in site using a Fluke cnx t3000 k type wireless thermocouple. Ardiuno Fio (described in Section 4.4) is used as a transmitter and the xbee module is used as the receiver. A dynamic robust structure is designed and developed to house the heat sinks. The structure has provision for the heat sinks to rotate and also to be oriented in any direction. The structure mainly consists of an arm, base and circular clamp, as shown in Figure 8.2.

The base of the structure is coupled with a stepper motor drive through a belt pulley assembly. A high torque motor (1000 rpm) is used to transmit the rotation to the structure. A microcontoller unit is designed and is programmed to ensure control of the voltage into the stepper motor to achieve various rotational speeds. The design of the structure is aimed at 3 point balancing to avoid the effect of vibration during the rotation of the heat sink at any given orientation. To ensure this, a vibration test is performed by using an accelerometer. The rotating cylindrical heat sink has an aspect ratio close to 1 (L/d = 50/48) and is made of aluminum. The bottom of the heat sink houses the plate heater.

8.2.1   Measurement of the cylinder surface temperature

For every set of experiments, surface temperatures are measured. The internal PCM/air temperature is not measured considering the movement of the solid/liquid phase for the cases with partially filled PCM. 28 gauge K type thermocouples are placed on the milled slots provided on the heat sink surface using Omega 201 bond. In view of the rotation of the apparatus, a wireless temperature circuit needs to be designed, made and deployed.

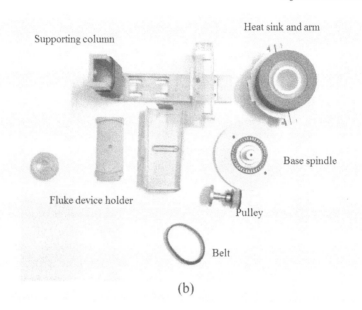

Supporting column

Heat sink and arm

Base spindle

Fluke device holder

Pulley

Belt

(b)

FIGURE 8.2
Photograph showing the components of the structure before assembly.

   The reference junction of the thermocouple is connected to a MAX 31855 thermocouple amplifier, which automatically takes care of the cold junction compensation. The signal from the thermocouple is of the order mV, in view of which, an amplifier is required. The amplified signal from the max 31855 (described in Section 4.4) is sent to the Arduino fio micro controller board. The Arduino fio is interfaced with the max31855 by a computer program. To the Arduino fio is connected an xbee transmitter. The digital signal with serial data transmission is done to the receiver. To regulate the voltage to the heater, the heater output is connected to LM 317 IC to which the battery input is connected. A LiPo battery of 11.V and 1800mAh is used. By using a variable voltage controller, the voltage to the heater is maintained constant. Initially, during the operation, it was observed that the constant current circuit was malfunctioning due to the self heating nature of LM 317 IC. To overcome this, an additional heat sink with a fan is used to maintain the IC temperature under limit during the full operation. The current output of the potentiometer is shown in Figure 8.3 for two cases, namely without and with the fan. The advantage of using a fan is clearly seen in the figure.

   In order to have accurate readings of the base temperature a Fluke cnx T 3000 wireless temperature module is employed. The fluctuations in the temperatures at high rotational speeds due to the effect of a magnetic field was also tested. The test was carried out by changing the polarity to the stepper motor(direction of rotation). It was seen that the temperature did not

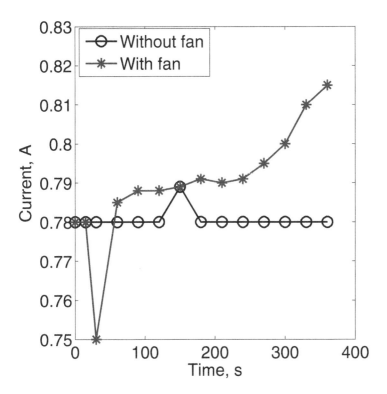

FIGURE 8.3
Effect of fan cooling on the output of potentiometer circuit.

change/fluctuate with the polarity and hence it was concluded that magnetic field effect is negligible. As the thermocouples are made of current carrying conductors, this test becomes mandatory. Furthermore, since the module is used to record the transient temperature history, the test is conducted for a longer period of time (3600s). The wireless measurement circuit is designed in-house and is calibrated using a constant temperature fluke calibration bath shown in Chapter 4. Calibration is performed during both the heating and cooling cycles. The results of the calibration are shown in Figure 8.4 and confirm the accuracy of the wireless temperature measurement process.

8.2.2 Measurement of rotational speed

A laser tachometer with an uncertainty of 1 rpm is used for the measurement of rotational speed. A reflecting sticker is pasted on the heat sink surface and the tachometer is focused on it to ensure one complete rotational cycle.

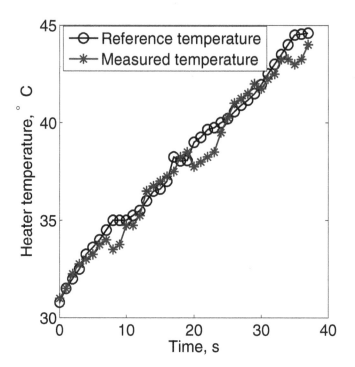

FIGURE 8.4
Result of the transient temperature calibration.

In the present work, the whole apparatus shown in Figure 8.5 is isolated from environmental noise. Experiments are conducted in an air conditioned room and throughout the experiments the ambient temperature was 26°C. Schematics of the orientations studied are shown in Figure 8.6.

8.3   Heat loss during experiments

To account for the heat loss that occur during the experiments, a numerical model mimicking experiments is developed using the enthalpy-porosity technique in Ansys Fluent 14.0, as discussed in detail in Chapter 5. For the sake of brevity, only the results obtained from the simulations are discussed. The numerical model was initially developed with an adiabatic boundary condition on all the outer surfaces, assuming no heat loss. Later the heat transfer loss coefficient was increased in steps of 0.5 W/m^2K. For an overall heat transfer coefficient of $4W/m^2K$ the numerical results (Temperatures at 5 different

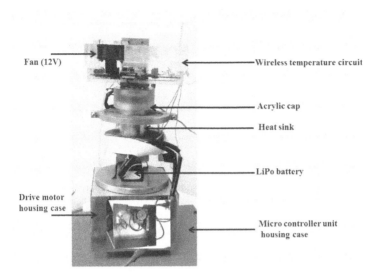

FIGURE 8.5
Photograph of the experimental setup.

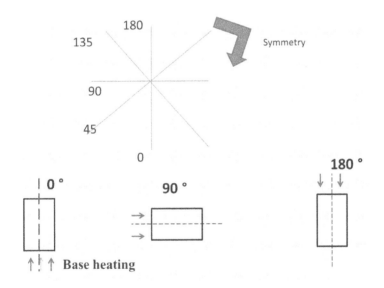

FIGURE 8.6
Schematics of various orientations investigated in the present study.

locations) showed good agreement with the experiments. From further calculations it was estimated that 0.3 W of heat is lost to the ambient for a total power level of 6W, which is within acceptable limits.

8.4   Results and discussion

8.4.1   Comparisons with the baseline

Leoni and Amon [44] stated that humans can comfortably handle plastic objects up to a temperature of 45° C. [29] conducted investigations on handheld objects and indicated that a temperature of 45°C is the threshold for thermal comfort. Based on these findings, it can be assumed that users will start to feel uncomfortable when the device temperature goes beyond 316 K. Thus, the set point temperature was set to be 316 K, in this study. Figure 8.7 shows the cases that have been considered for both the melting and the solidification cycles.

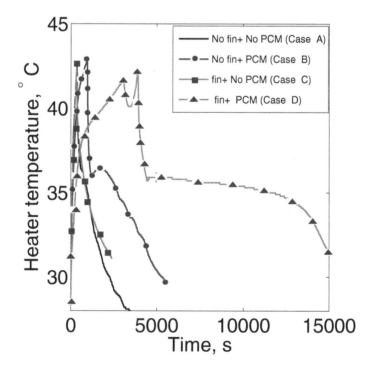

FIGURE 8.7
Temperature time history for 4 key situations (A - No fin and no PCM, B - No fin with PCM, C- Fin with no PCM, D - Fin with PCM).

In all the cases considered, the volume of the PCM used remains constant. Case A represents an empty heat sink (air), with predominantly sensible heating. Case B (cylinder with no fins and filled with PCM) represents the effect of PCM to store the latent heat, which is a victim of the "self-insulating" effect, while case C (cylinder with internal fins and no PCM) represents the effect of the fin in dissipating the heat. Case D (cylinder with internal fins and PCM) represents the combined effects of latent and sensible heat. From the experiments, it is evident that case D is superior to all other cases in maintaining the device temperature for a longer period of time. Case D reiterates the fact that in order to make use of the latent heat of PCM, the heat from the base needs to be spread uniformly in the PCM. Among the other cases, case C with only PCM performs better due to the lower rate of change of temperature. The superior performance of case D results from the effective utilisation of latent heat, by overcoming the low thermal conductivity of PCM. As expected, the cooling time is very high for case D. Figure 8.7 establishes the basic trade-offs that we are trying to resolve throughout this book.

8.4.2   Thermal performance of the unfinned heat sink

The unfinned heat sink is subject to simultaneous variation of fill ratio and orientation. Throughout this study, the fill ratio translates to volume percentage of PCM inside the cavity. Figure 8.8 shows the performance of the heat sink as a function of orientation for four fill ratios namely 0, 0.33, 0.66, 0.99 for a heating of 6W at the base. The figure clearly establishes that both the parameters, fill ratio and orientation exert a strong influence on the thermal performance. Though it is a ratio, in this study it is multiplied by 100 to get a % which is more convenient and easy to comprehend. From Figure 8.8, it is seen that for 0° orientation, the 99% fill ratio case operates for a longer time during the charging cycle. To better evaluate the effectiveness of using a PCM, the enhancement ratio (ER), defined earlier is used.

   This ratio can also be used to evaluate the performance between various fill ratios of the PCM. From Figure 8.8 one can see that at 0° orientation, the enhancement ratio increases all the way up to a fill ratio of 99%, though the rate of this enhancement decreases with this fill ratio. This is a consequence of the self-insulating nature of the PCM. Additionally, for a PCM-filled heat sink, two peaks at 90° and 270° are seen at intermediate fill ratios, while the effect of orientation is muted for a fill ratio of 99%. It is quite possible that in actual applications, the orientation of the heat sink can change with time in an actual application. This is the major motivation behind the study of the heat sink performance at various orientations. The striking obsevation from Figure 8.8 is that the 99% fill ratio is insensitive to orientation which is a delight to the engineer, for reasons articulated a little earlier.

   When the volume fraction of the PCM is 0.33, the heat sink performance shows a sudden enhancement at an orientation of 90°. The partially filled heat

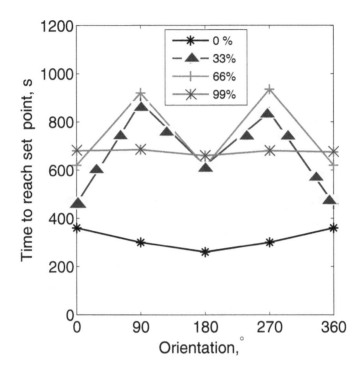

FIGURE 8.8
Performance of the unfinned heat sink at different fill ratios and orientation
for a power level of 6W at the base.

sink is able to utilise the latent heat effectively without the aid of TCEs by
the phenomenon of "wall cooling", for this orientation.

To understand the wall cooling effect, the experiment was stopped at
intermediate points and photographs as shown in Figure 8.9 at various time
instances are taken. The photographs are also captured for the vertical orientation (0° case) and these are shown in Figure 8.10. Both Figures 8.9 and
8.10 are for a partially filled heat sink with the fill ratio being 33%.

As the photographs in Figure 8.9 indicate, initially the PCM gets heated
from the base and the side walls. Due to gravity, the PCM layer close to
the upper side wall melts and slides down. As the solid PCM slides, more
of the PCM is now exposed to the wall leading to more PCM melting. Due
to gradual melting of more PCM mass, the heat sink temperature is under
control for a very long time. This reiterates the fact that for an unfinned heat
sink, complete filling is not the optimal solution. Furthermore, a better way
is to partially fill and orient it at 90°. In fact most of the portable electronic
devices are used at 90°. At the end of the melting, one can observe that the
complete mass of the PCM has melted. The latent heat content can then be

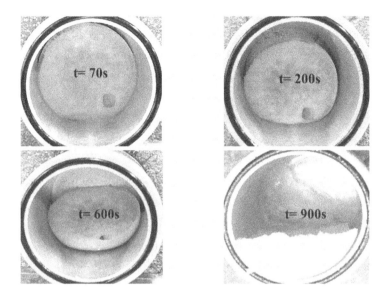

FIGURE 8.9
Photograph showing the melting pattern in 33% filled unfinned heat sink at 90°.

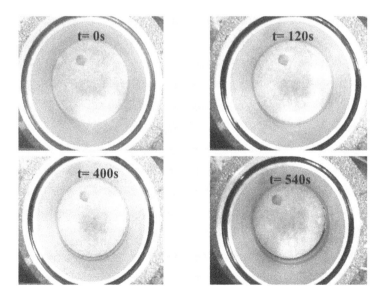

FIGURE 8.10
Photograph showing the melting pattern in 33% filled unfinned heat sink at 0°.

TABLE 8.1

Latent heat content for different fill ratios for an orientation of 90°

Fill ratio (%)	Latent heat content (kJ)
33	3.98
66	7.97
99	11.96

calculated as the product of the PCM mass and the latent heat of the PCM. Results are shown in Table 8.1.

For the 33% fill ratio case, as the photographs in Figure 8.10 indicate, the 0° orientation is able to aid only 20%(approximately) of the PCM mass to melt. The latent heat utilised in this case is 0.8 kJ. However, at 90°, the entire mass of the PCM has melted indicating complete utilisation of latent heat content. The thermal performance of unfinned heat sink at 0° and 90° is shown in Figure 8.11. From the figure, it is clear that the wall cooling effect

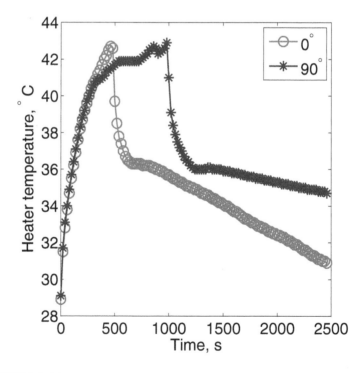

FIGURE 8.11

Comparison of temperature time history of the temperature at the base for an unfinned heat sink with 33% fill ratio at 0° and 90°.

extends the latent heat time for a 90° orientation. However, the solidification time is also higher for a 90° oriented heat sink.

However, the penalty for the complete melting of PCM is seen in the solidification cycle. Invariably, it takes a longer time for the heat sink to return to the initial state. This effect of wall cooling is observed from 90° to 97°. The complete utilisation of the latent heat in the 90° case during melting results in latent heat release and hence the heat sink takes a longer time to solidify.

After 110°, the melted solid detaches itself from the base and slides away, allowing the air to occupy the region resulting in a more insulating effect which leads to a high temperature rise. When the orientation is changed from 110° to 180° (heating from above), regardless of the fill ratio, all the cases perform nearly the same due to the sudden detachment of the solid PCM from the heated base. Among all the orientations, it is evident that 180° orientation does not allow for the complete melting of PCM and there is a sudden change in the position of volume occupancy that leads to a steep temperature rise. As the set point temperature increases, the performance of the heat sink at 180° is worse still. The solidification time is also very short due to the presence of a high amount of unmelted PCM (mass that did not absorb latent heat). For a PCM volume fraction of 0.66 too, the same trend is observed for all the orientations. This effect reduces as the fill ratio increases. The additional mass of PCM does not ensure good performance at all orientations.

8.4.2.1 Effect of rotation on the thermal performance of unfinned heat sink

For an air based unfinned heat sink (0% fill ratio), the effect of rotation is found to be negligile during the heating cycle. However, the cooling cycle time is reduced considerably due to the rotation of the heat sink at 120 rpm as shown in Figure 8.12. It is evident that the rotation of the heat sink induces a forced convection at the external surface of the heat sink. This effect is predominant for all orientations of the heat sink. Stagnant air near the heat sink wall can be possibly overcome by rotating the heat sink. The cooling time also varies with the orientation of the heat sink. The effect of rotation was also tested for a PCM-based heat sink. Tests were conducted for all fill ratios and orientations. The performance of the heat sink enhances by 25% during the melting process. For a fill ratio of 33% and 0° orientation, the time to reach set point at the heated base increases by 25% as shown in Figure 8.13. It is evident from the temperature time history that there is a strong deviation of temperature only when a certain portion of the PCM is melted (post latent heat region in the temperature history). Additionally, the Nusselt number curve remains constant against the product of Fourier number and Stefan number indicating additional advection of liquid PCM, thus extending the operating time.

Counterintuitively and surprisingly, the solidification time is not improved significantly due to rotation. This reiterates the fact that the only mode of heat

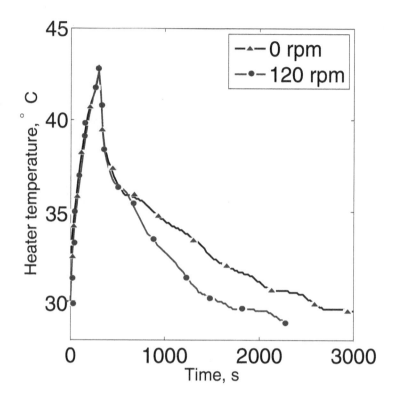

FIGURE 8.12
Performance of unfinned air based heat sink under rotation.

transfer during solidification is conduction and external wall convection too is negligible due to the high specific heat capacity of the PCM. A much higher rotational speed is possibly required to induce an external forced convection due to rotation. As the fill ratio is increased, the effect of rotation is found to be negligible.

8.4.3 Thermal performance of the heat sink with a central stem

This particular configuration of heat sink is studied with a view to investigate the effect of spreading of TCEs on the melting and solidification process. The volume of the stem is equal to the total volume of stem and the fins in the finned heat sink case. This study can quantify the effect of the stem on the melting and solidification process clearly. Figure 8.14 shows the heat sink performance during melting as a function of fill ratio and orientation, for the same power level of 6W as was applied for the heat sink with no

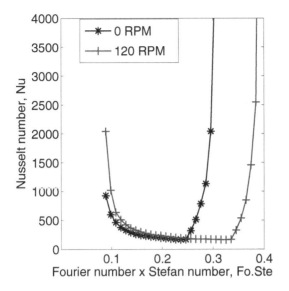

FIGURE 8.13
Comparison of time history of base Nusselt number at the base for an unfinned rotating heat sink with 33% fill ratio under rotation for a power level of 6W at the base.

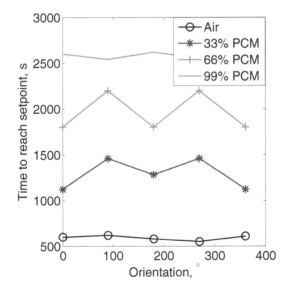

FIGURE 8.14
Performance of heat sink with a central stem at different fill ratios and orientation for a power level of 6W at the base.

fins in Section 8.4.2. For the 99% fill ratio case, the heat sink takes 2600s
to reach the setpoint during melting and this value is more or less for all
orientations indicating that the heat sink with 99% fill ratio is again robust.
The performance of the heat sink is between that of the no fin and finned
heat sink case. Experiments clearly indicate that the internal vertical stem is
able to reduce the temperature gradient in the Y direction. However, in the X
and Z directions, the thermal resistance is high. Furthermore, the orientation
of the heat sink affects the heat transfer performance much in the same way
as that of the unfinned heat sink. From the experimental observations, it is
clear that melting predominantly occurs near the wall and the external stem
surface while the stem is able to reduce the temperature gradient to some
extent. Even so, this effect is negligible at higher fill ratios.

8.4.4   Thermal performance of the finned heat sink

Experiments are then conducted for the same power of 6W at the base as in
Sections 8.4.2 and 8.4.3. The finned heat sink is first tested with a fill ratio
of 0%. These experiments showed that the heat removal is associated with
steep temperature rise (sensible heating). The heat sink is then subject to
simultaneous variation of fill ratio and orientation and the results are shown
in Figure 8.16. As the fill ratio is increased, better performance is achieved.
An enhancement ratio, of 10.25 is achieved at 99% fill ratio.

For a fill ratio of 33%, as the orientation is increased from 0 to 180°, the
thermal performance is seen to monotonically decrease, as shown in Figure
8.15. Because of the presence of TCEs, 0° orientation is able to promote solid
sinking and complete melting of PCM is achieved.

Figure 8.16 shows the thermal performance of the finned heat sink during
melting as a function of fill ratio and orientation.

It is seen from Figure 8.16 that the thermal performance varies monotonically with fill ratio and orientation. It can easily be concluded that the heat
sink performs the best at 99% fill ratio and 0° orientation. Furthermore, at
180°, there is an early detachment of PCM, leading to a quick temperature
rise at the heat sink base. However, as the fill ratio is increased from 0 to 99%,
the effect of orientation is negligible.

From Section 8.4, it is clear that the finned heat sink is superior to the
unfinned and central stem heat sink. This section presents the results of the
finned heat sink which is subject to rotation at all fill ratios and orientations
simultaneously. At lower fill ratios(33%), rotation is able to enhance the time
of melting as shown in Figure 8.17 and extend the operation time of the heat
sink.

Figures 8.18a show the thermal performance of the finned heat sink with
66% and 99% fill ratio PCM respectively.

At 66% fill ratio too, the latent heat time is extended as shown in Figure
8.18a. However, at 99% fill ratio, the effect of rotation is negligible, as is clearly
seen in Figure 8.18b.

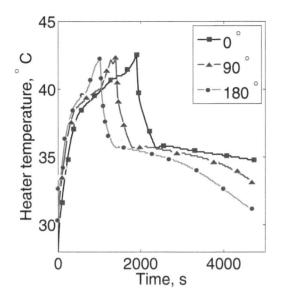

FIGURE 8.15
Performance of finned rotating heat sink at different orientations at 33% fill
ratio for a power level of 6W at the base.

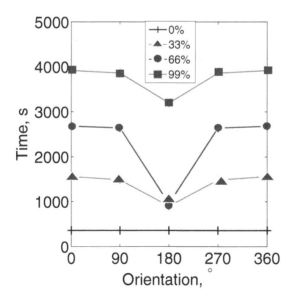

FIGURE 8.16
Performance of heat sink with fins at different fill ratios and orientation for a
power level of 6W at the base.

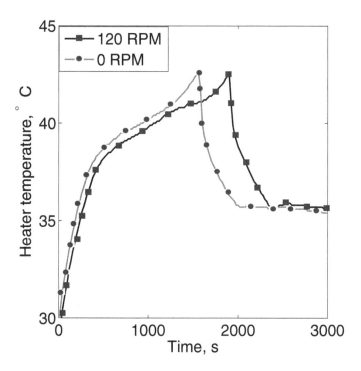

FIGURE 8.17
Comparison of temperature time history of the temperature at the base for a
finned rotating heat sink with 33% fill ratio under rotation.

8.5   Numerical analysis

As a precursor to the optimization studies limited exploration of full numerical
simulation is attempted. Since the effect of rotation was already discussed
through experiments, numerical investigation is done to better understand the
effect of geometry on the performance of the heat sink for the no rotation cases.
The volume of TCEs and PCM remains constant in the cases investigated.
However, the distribution of TCEs differs. A vertical cylinder with a central
stem with radially outward fins is the geometry under consideration, as shown
in Figure 8.19.

The cylinder contains PCM and the heat sink material is aluminum as
before. The PCM is n-eicosane. Different heat sink configurations are considered with a constant wall thickness of 5mm. At time t=0, temperature of
300K is assumed everywhere. The enthalpy porosity method, as discussed in
Chapter 5 is employed here as well.

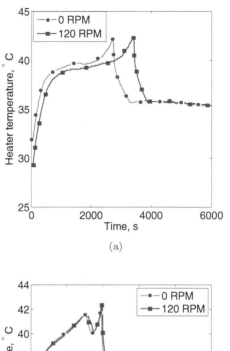

(a)

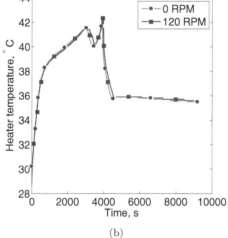

(b)

FIGURE 8.18
Comparison of temperature time history of the temperature at the base for a finned rotating heat sink with(a) 66% fill ratio under rotation (b) P99% fill ratio under rotation.

Calculations are done for 40 different cases using ANSYS fluent 14.0. From the results of grid independence studies, a grid with 745,000 nodes was seen to be adequate. Furthermore, from the results of time step independence studies a time step of 0.05s was seen to give time step independent results. The convergence criteria for the mass, momentum, energy equations are 10^{-6}, 10^{-6} and 10^{-12} respectively. The VIRGO super cluster at IIT Madras was

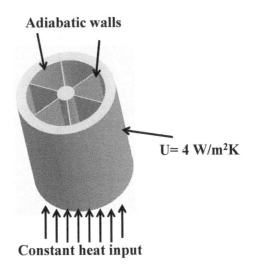

Adiabatic walls

U= 4 W/m²K

Constant heat input

FIGURE 8.19
Picture showing the boundary conditions on the surface of the heat sink model.

employed for the numerical simulations. Among the 40 cases, 4 different cases are chosen and their geometric details and corresponding time for complete melting are given in Table 8.2.

The parameters d_i, h_i, d_s, t_f and n are significantly different for all the cases. Among the 4 cases shown in Table 8.2, cases 5 and 18 are taken up for a detailed investigation of the fluid flow and heat transfer characteristics. The volume of TCE and PCM remains the same in all the cases. The heat input is constant at the base the same as the experiments.

The geometric difference between the two cases from Table 8.2 are stated below.

(i) Case 18 has only 2 fins, whereas case 5 has 3 fins.
(ii) Fins in case 5 are 0.4mm thicker than that in case 18.
(iii) Case 18 is 7.5mm taller than case 5.

TABLE 8.2
Details of some of the geometries investigated

Case	di, mm	hi, mm	ds. mm	t, mm	n	t melt, s
5	40	51	8.5	1.65	3	4200
6	43	43	6.8	0.6	8	3529
18	37	58.5	9.6	1.27	2	3400
26	46	39.2	8.7	0.5	11	3570

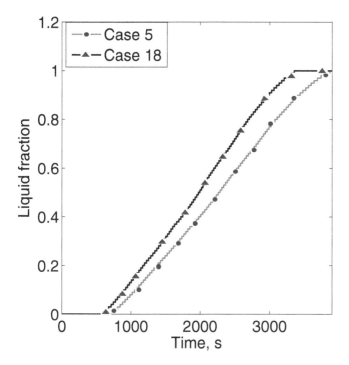

FIGURE 8.20
Variation of liquid fraction with time for cases 5 and 18.

Figure 8.20 shows the transient liquid fraction for two cases. A liquid fraction of 0 corresponds to solid PCM and 1 corresponds to fully molten PCM. It is intuitive to note that the liquid fraction presented here is a volume averaged quantity, with the averaging done all over the cell volumes. From Figure 8.20, it is clear that in case 18, the time to fully melt is 500s earlier than for case 5.

For the purpose of heat transfer analysis, the vertical temperature distribution on the wall of the stem is calculated for every 500s. From the geometry, it is very clear that the PCM layer thickness is higher for case 18 in a sector. Sector analysis is considered in this study because the heat from the base spreads symmetrically in each sector. The initial mode of heat transfer is conduction. As the melting of the PCM begins, convection currents begin to develop as a result of density variation. After 1000s of heating, there is a temperature gradient of 23.2 K/m set up along the stem. As time progresses the temperature gradient keeps on decreasing.. At t=1000s, the slope is very high. During this process, the PCM layer adjacent to the stem and the fin wall begins to melt. Since the density of the PCM is a function of temperature, this temperature gradient induces a density difference and the fluid circulation begins. The strength of the convection currents influences the molten PCM shape and the duration of complete melting. As time progresses, the velocity of the liquid PCM keeps increasing up to t=2500s, when 71% of the PCM is molten. At t=3000s, the velocity of the fluid drops considerably. This stage corresponds to a PCM melt fraction of 90%. The stem is almost isothermal with a low temperature gradient of 9.64 K/m. As the liquid PCM approaches full melting, a sudden increase in the velocity is observed. This phase of the process is purely dominated by convection during which both the velocity and temperature of the PCM keep increasing steadily.

The heat transfer to the PCM comprises three components, namely

- Heat transfer directly from the base

- Heat transfer from the stem

- Heat transfer from the fins.

8.5.1   Analysis of a specific geometry (case 5)

In case 5, there are three radial fins present. This implies that each sector has a lower PCM width compared to case 18. Furthermore, case 18 is taller than case 5. The temperature profile on the wall of the vertical stem is observed for every 500s. At any time instant, anywhere on the stem the temperature is lower in case 5 than compared to that in case 18. The temperature gradient on the stem wall keeps on decreasing until t=3000s. At this instant of time only 77% of the PCM is molten. This is not the situation in case 18, where at this instant of time 95% of the PCM is molten. After 3000s, there is a sudden drop in the liquid PCM velocity and a slight increase in temperature gradient too.

From t=2000s to t=3500s the PCM temperature remains isothermal. The velocities are observed to be almost constant from t=2500 to 3500s. The relatively constant velocity observed during the melting in case 5, contributes to the consistent heating of the solid PCM. The temperature gradients in the radial direction are also calculated. For case 5, at t=2000s, the temperature near the stem wall is 312.5K, in the mid plane it is around 309.1K and near the heat sink wall it is 310.6K. This clearly indicates that the heat transfer into the PCM from the stem is much higher than from other sources. The radial gradient of temperature from the stem to mid plane is found to be 350K/m which is much higher than the temperature gradient in the vertical direction at the same time instant. The liquid fraction keeps on increasing with time for both the cases.

8.5.2   Analysis of centre line temperatures for cases 5 and 18

A study of the centre line temperature helps one understand how fast the heat diffuses into the centre of the heat sink. For both cases the centre line of the sector is chosen for the study. Figure 8.21 shows the centre line temperature

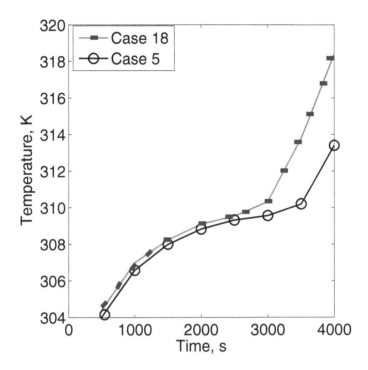

FIGURE 8.21
Temperature time history at the centre line for cases 5 and 18 (Refer to Table 8.2 for the details of case 5 and 18).

history for the two cases. At any instant of time the case 18 shows a higher temperature at the centre of the heat sink. During the initial stages of melting there is not much difference in the temperatures between the two cases and this trend continues until 2500s.

However, beyond 3000s there is a sudden increase in temperature. This indicates that the convection has become strong and dominant in case 18 which erodes the solid PCM and results in a quick rise in temperature. At the end of 4000s the centre temperature in case 18 is 5K higher than case 5. This analysis is useful in understanding the importance of convection.

Figure 8.22 depicts the effect of Stefan number on the average Nusselt number on the heated base for case 5. The scaled Nusselt number is on the ordinate and the product of Fourier number and Stefan number is on the abscissa. This plot is specific to case 5.

The scaled Nusselt number([31]) decreases during the initial stages of melting. This is a well-known characteristic of transient heat conduction. This happens during the initial stages of melting, where the primary mode of heat transfer is conduction. Later, when buoyancy driven convection currents are induced in the flow, the decrease in heat transfer coefficient is slowed down

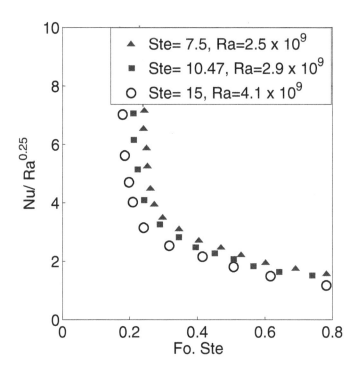

FIGURE 8.22
Effect of Stefan number on the melting process for case 5 (Refer to Table 8.2 for the details of case 5).

and it reaches a quasi-steady value and remains constant thereafter. This corresponds to the heat transfer to a completely melted PCM.

As the Stefan number is increased, the upper portion of the curve tends to shift to the left indicating that melting is initiated at early values of (Fo.Ste). At later stages, the curves converge to a single point. Additionally, the increase in Stefan number directly influences the time for the complete PCM to melt in both the cases. From Figure 8.22, it is also clear that for higher Rayleigh numbers, the quasi steady state heat transfer coefficient is reached at later values of (Fo.Ste)

8.6 Engineering perspective of the cylindrical heat sink configurations

A bird's eye view of the matrix of configurations tested in the present study is shown schematically in Figure 8.23. From an engineering standpoint, the two main objectives are (i) low cost and (ii) high performance. These, unfor-

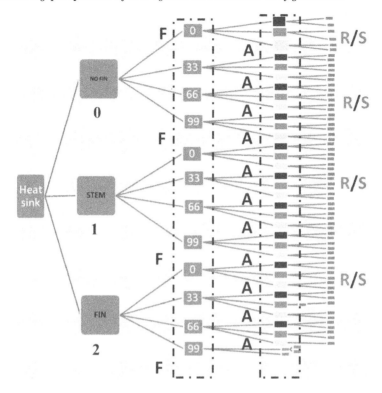

FIGURE 8.23
A bird's eye view of all the configurations investigated in the present study.

tunately, are in conflict with each other. The performance and cost of each configuration are rated on a scale of 1 to 10. The performance of the heat sink during heating/melting is defined as the ratio of the time to reach set point of the configuration to that of the highest time among all the configurations. The performance of the heat sink during cooling/solidification is the ratio of the time to reach set point of the configuration to that of the least time among all configurations. The cost consists of the fin cost, PCM cost and cost of rotation. A scatter plot of these two figures of merit for a few randomly chosen configurations is shown in Figures 8.24 and 8.25 for the cases of melting and solidification respectively. The '+' sign indicates the ideal solution and the '-' sign indicates the undesirable solution. The spread of the data points can be categorised into four quadrants, Q_1, Q_2, Q_3, Q_4. Q_1 represents the heat sink configurations with high cost and low performance and so on. From an engineering standpoint, Q3 is the most desired quadrant. From Figure 8.24 it is seen that the most ideally desired configuration does not exist. Strangely, though the most undesirable is easy to spot. When melting heat transfer alone is considered, quadrant 1 is highly undesirable. The nomen-

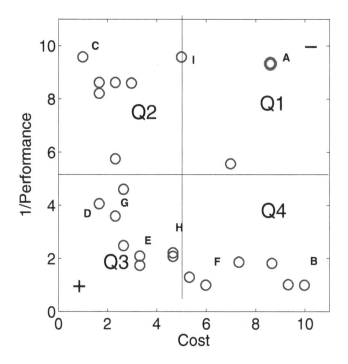

FIGURE 8.24
Figure showing the cost and the performance metrics of a few random configurations during melting.

clature 0-33F-90A-R [Fin type- (Fill ratio in %) F- (Orientation angle) A-
Rotation/Stationary] can be used to qualify the heat sink which in this case
turns out to be an unfinned heat sink with 33% fill ratio oriented at 90° and
that is undergoing rotation.

From the analysis, it is evident that most of the cases that perform well
during melting do not perform well in solidification. This is seen from Figures
8.24 and 8.25. The two represent the trade off between cost and performance
for any configuration for melting and solidification respectively. The scatter is
divided into four quadrants, namely Q1, Q2, Q3, Q4. Q1 represents the highest cost and low performance which is undesirable. Similarly, Q3 represents
best performance at low cost which is highly desirable. The '+' indicates the
most desirable point(which ideally does not exist) and '-' represents the most
undesirable point (which exists).

One can infer that no configuration holds the same position for melting
and solidification. Configuration A (Refer to Table 8.3) which was a part of
quadrant 1 in melting (Figure 8.24), is midway between quadrant 3 and 4
Figure 8.25. Also, Configuration E (Refer to Table 8.3) which was a part

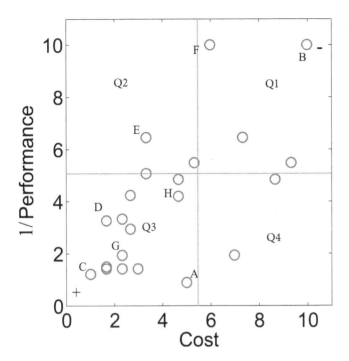

FIGURE 8.25
Figure showing the cost and performance metrics of a few random configurations during solidification.

of quadrant 3 in melting, occupied quadrant 1 in solidification. Furthermore, Configuration B (Refer to Table 8.3) which was a part of quadrant 4 in melting, turns out to be the most undesirable configuration in solidification. This shift in configurations reiterates the fact that melting and solidification are highly conflicting objectives, necessitating the use of robust optimization to resolve the trade-offs.

Additionally, some cases are more costly and are worse off in performance. 1 indicates good perfomance and low cost. 0 represents the opposite.

Now we consider only melting heat transfer as shown in Figure 8.24. Case D shows good performance and reasonable cost. A higher order information is now required to choose the best configuration one desires. For instance, if the premium is on performance and not on the cost, then case B is the best. If equal weightage is given to cost and performance then case E can be chosen. This happens to be the case with the central stem. The diverse spread of the initial samples can pave the way for future multi-objective optimization studies involving both cost and performance for melting and solidification considered individually and then together.

TABLE 8.3
Table showing different configurations with their respective performance and
cost metrics

Case ID	Configuration	Performance		Cost [b]
		Melting	Solidification	
A	0-0F-0A-R	0	1	0
B	2-99F-0A-R	1	0	0
C	0-0F-0A-S	0	1	1
D	0-33F-90A-S	1	1	1
E	1-66F-90A-S	1	0	1
F	2-99F-0A-S	1	0	0
G	0-66F-0A-S	0	1	1
H	2-33F-90A-S	1	1	1
I	0-0F-0A-R	0	1	0.5

8.7 Conclusions

Heat transfer experiments and numerical studies were conducted for cylindrical heat sinks with three configurations namely

- Unfinned heat sink

- Heat sink with an internal stem

- Heat sink with stem and fins

The heat sinks were subject to different fill ratios, orientations and rotation speed simultaneously. Based on detailed parametric studies, the following
important conclusions are arrived at

- The performance of PCM-based heat sink is superior to air-based heat
  sink during the heating/melting cycle, and vice versa during the cool-
  ing/solidification cycle

- For an unfinned heat sink, a higher fill ratio does not ensure superior
  performance at all orientations

- Keeping the heat sink at 180° proves fatal during the melting regardless
  of any configuration

- A finned heat sink performs the best among the three chosen configurations
  with an enhancement ratio of 10.25, while incurring the highest cost.

- For an air-based heat sink, rotation enhances the cooling cycle time by
  30%

- For a finned heat sink, the more the PCM, the more superior is the performance.

- Regardless of the configuration, rotation has a positive effect of 25% performance enhancement on melting, only at lower fill ratios. However, rotation is found to have a negligible effect at higher fill ratios.

- The orientation of the heat sink plays an important role for an unfinned heat sink. The performance of the heat sink during melting is not monotonic with the orientation.

- For a finned heat sink, the heat sink performance is a strong function of number of fins, fin thickness and stem diameter. Different combinations of these variables yield a different thermal performance.

In sum, to conclude there is always a conflict between the cost and the performance for all configurations. This can be resolved by the designer keeping in mind the application and the requirement. The efforts made in this study clearly pave the way for future multi-objective optimization studies when the heat sink is used in actual applications.

8.8 Closure

The present chapter discussed the heat transfer in a cylindrical heat sink subject to different orientation, fill ratios and rotation simultaneously. The next chapter discusses a synergistic numerical approach to solve thermosyphon-assisted melting of PCM.

9

THERMOSYPHON ASSISTED MELTING OF PCM INSIDE A RECTANGULAR ENCLOSURE: A SYNERGISTIC NUMERICAL APPROACH

9.1 Introduction

Latent heat thermal energy storage systems (LHTES) that employ Phase change materials (PCMs) are considered to be one of the most efficient techniques to store energy. The main advantage of using a LHTES is its ability to store thermal energy with a minimal temperature change. However, the low thermal conductivity of PCMs poses a great challenge to heat transfer experts owing to the difficulty in heat transport. Various techniques like fins and heat pipe have been adopted in conjunction with PCMs to enhance the thermal conductivity of the LHTES over the years. The idea of using fins, rotation and so on has been elaborately presented in the earlier chapters. A thermosyphon is a gravity assisted wickless heat pipe containing a working fluid, which undergoes phase change during the transient operation of the heat pipe. The evaporator section of the heat pipe filled with water absorbs the major portion(>80%) of the heat input. The condensation at the condenser serves as the heat source for the PCM. In this chapter, results of investigation with heat pipes being used as TCE are presented based on the work by [68].

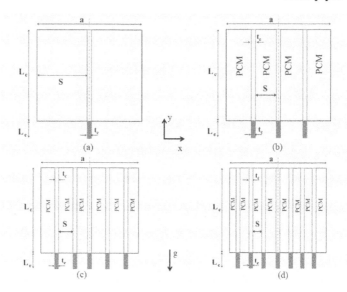

FIGURE 9.1
Models of heat pipe-PCM system studied [68].

TABLE 9.1
Model description [68]

Model No	Number of HPs	Spacing (S), mm
1	1	27.5
2	3	11.25
3	5	5.75
4	7	3.125
5	9	1.5

9.2 Physical model

A schematic view of the geometries investigated is shown in Figure 9.1. Five sets of geometries, with single and multiple heat pipes are modeled.

The heat pipes comprise an evaporator section and a condenser section and are made of copper. The length(L_e) and diameter (d_e) of the evaporator section are 30 and 50 mm respectively. The length(L_c) and diameter(d_c) of the condenser section are 60 and 50 mm respectively. The details of the cases studied are tabulated in Table 9.1.

The entire model is at an initial temperature of 300 K. The evaporator section of the heatpipe is supplied with a total heat input of 6 W. In model 1, the system is designed in such a way that the entire heat load is taken up by a central heat pipe and aids in the melting of PCM. In model 2, the system is designed in a way that the total heat load is shared by three heat pipes that are equally spaced. In model 3, 4 and 5, the heat load is shared by five, seven and nine heat pipes respectively. Hence, the addition of heat pipes essentially redistributes the total heat input.

9.3 Numerical procedure

9.3.1 PCM

The rectangular domain contains PCM and the heat pipe (HP) unit. The PCM used in this study is n-eicosane. The properties of the PCM and the governing equations are the same as stated in Chapter 3 and Chapter 7 respectively in detail.

9.3.2 Heat pipe

Throughout this chapter, heat pipes only mean thermosyphons. The operation of the wickless heatpipe is modeled using a simplified lumped parameter model. Figure 9.2 explains the nodalization of the lumped model.

The following assumptions are used for the lumped modeling of wickless heatpipe:

- One dimensional heat transfer

- Uniform heating and cooling

- Vapor and liquid inside the thermosyphon are at thermal equilibrium and hence at saturation state

- No axial heat losses

- Fluid thermal capacity is approximated as the liquid thermal capacity since vapour thermal capacity is negligible

The governing equations for the model are as follows

$$C_w \frac{\partial T_w}{\partial t} = Q_e - h_e S_e (T_w - T_f) \tag{9.1}$$

$$C_f \frac{\partial T_f}{\partial t} = h_e S_e (T_w - T_f) - h_c S_c (T_f - T_{wat}) \tag{9.2}$$

$$T_{cond} = T_{wat} + \frac{Q_{net}}{h_c \times A_c} \tag{9.3}$$

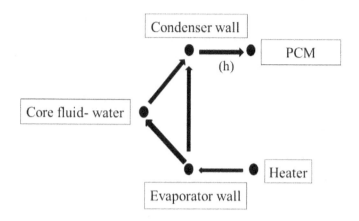

FIGURE 9.2
Schematic of coupling procedure [68].

In this study h_e is the average evaporator heat transfer coefficient. The nucleate boiling heat transfer coefficient is calculated from the Froster-Zuber equation.

9.3.3   Coupling

The lumped model discussed above is incorporated with ANSYS fluent as User Defined Function (UDF) subroutines. The condenser flux calculated by the UDF is applied as the input flux for the condenser section in the Enthalpy porosity model.

The Nusselt number from the heated wall is calculated by using the following relation

$$Nu = \frac{h_c \times L_c}{k} \tag{9.4}$$

To account for the buoyancy effect inside the liquid PCM that drives the heat transfer, the Rayleigh number is included in the study. The Rayleigh number is also defined [6] based on the PCM thickness.

$$Ra = \frac{g \ \beta \ \triangle T \ L_c^3}{\alpha \ \nu} \tag{9.5}$$

The heat transfer coefficient(h_c) at the condenser wall-PCM liquid interface is calculated based on a quasi-steady approach such that during each step the system proceeds through a sequence of quasi-steady states for dynamic equilibrium. The case of natural convection from a vertical wall under constant heat flux at every time instant is used to calculate h_c. From the iso-heatflux condition on a vertical wall, the Nusselt and Rayleigh number can be correlated by using the wall averaged Nusselt number relation mentioned in Chapter 5 of [12].

9.4 Validation

The enthalpy-porosity model and the lumped heat pipe model has been validated by experimental results available in the literature. Furthermore, the heat pipe lumped model is validated with the experimental results of [26]. as shown in Figure 9.3.

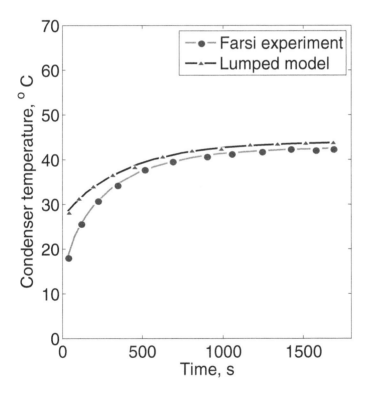

FIGURE 9.3
Validation of the lumped model with Farsi et al.[2003] [68].

Commercially available ANSYS fluent 14.0 is used to solve the governing equations. Due to symmetry of temperature and flow fields with respect to y axis, the domain is modeled as symmetry; for each case, only one half of the entire geometry is considered with a symmetry boundary condition at the centrally placed heat pipe axis.

Grid independence results indicated that a total of 100456 and 56768 elements was found sufficient for the 1 heat pipe case and 7 heat pipe case respectively. A time step size of 0.05s was found sufficient as additional time size refinement did not enhance the accuracy of the computation. In order to improve the convergence stability in this study during the phase change process, the under relaxation factors of 0.5, 0.3 and 0.9 were considered for the momentum, pressure and melt fraction, respectively. Convergence residual values of 10^{-6}, 10^{-6} and 10^{-10} were set for continuity, momentum (x and y velocities) and energy equations, respectively. In each time step, the convergence criteria were achieved after 150 iterations.

9.5   Results and discussion

To evaluate the performance of LHTES, the following parameters and their transient variation are considered in this study

- condenser wall temperature

- PCM melt fraction

- liquid PCM velocity

- condenser wall heat flux

For model 1 with single HP, there is a large thermal resistance setup between the heated wall(condenser) and the unmelted PCM zone. This results in high temperature of both the condenser and the evaporator wall. Initially, the melt fraction of the PCM increases with time. The rate of melting gradually decreases with time when 5% of the PCM is melted. This is due to the fact that the PCM closer to the wall takes up all the thermal energy and forms a zone of high thermal resistance for the flow of heat.

The LHTES thermal conductivity can be enhanced by incorporating more HPs in the area experiencing high resistance for the flow of heat. Addition of HPs leads to a decrease in PCM mass in the system. Hence, an optimal value of HPs to the PCM in a LHTES system is always desired.

In models 2 and 3, as the number of HPs is increased (spacing is decreased) the charging process becomes fast as seen in Figure 9.4. In cases with lower spacing, the natural convection in the liquid PCM near the heated surface helps in transferring heat to the unmelted PCM zone. Hence, the growth of

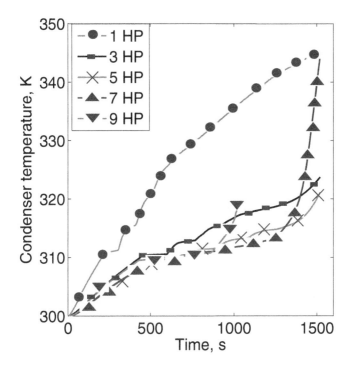

FIGURE 9.4
Transient variation of condenser temperature for the models considered [68].

wall temperature is much slower in models 2 and 3. From this it is evident that the natural convection plays a significant role in the melting process.

The volume ratio v_h of the HPs can be defined as

$$v_h = \frac{Amount \ of \ TCE}{Amount \ of \ PCM} \tag{9.6}$$

Furthermore as the number of HPs is increased, the mass of the PCM in the system is reduced proportionally. It is intuitive that there exists an optimum ratio of HPs to the mass of the PCM for the best performance of LHTES.

For cases with 3 and 5 HPs, the LHTES system provides high melt fraction and lower base wall temperature as seen in Figure 9.4. This is a direct effect of the natural convection where the PCM layers receive heat from both the boundaries.

The condenser wall temperature is calculated throughout the length of the condenser. The PCM velocity and mass fraction are calculated in the PCM module. The volume averaged parameters are taken for the PCM portion.

Furthermore, from the transient history of melt fraction seen in Figure 9.5 for the four cases, it is evident that as the number of heat pipes is increased,

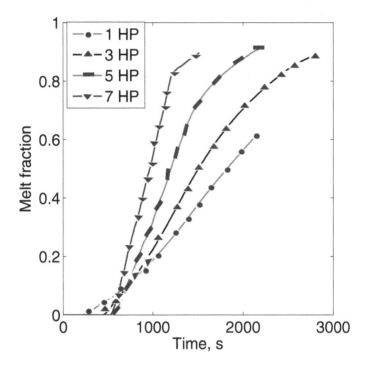

FIGURE 9.5
Transient variation of melt fraction for the models considered [68].

the time to melt 70% of PCM decreases monotonically. This is invariably due
to the fact that the increase in HPs results in decrease of PCM mass. The
transient variation of liquid PCM velocity and condenser wall heat flux are
shown in Figure 9.6 and 9.7 respectively.

To evaluate the performance of the LHTES, the condenser wall temperature after 1500s of heating is monitored. The lower the wall temperature,
the better is the performance. As can be seen from Figure 9.4, the single HP
case has the maximum temperature of 344K after 1500s of charging cycle. As
the number of HPs is increased, the performance of LHTES systems becomes
better.

For the 5 HP case the system is maintained at 319 K. An increase in heat
pipe number beyond this degrades the performance of the LHTES system
(Indicated in bold face in Table 9.2). The temperature of the condenser wall
shoots up to 330K. Hence, caution needs to be exercised while designing the
LHTES system to keep an optimum ratio of the volume HPs to the PCM to
enhance the system performance.

However, addition of HPs beyond a particular number has an adverse effect
on the LHTES. Furthermore, from Table 9.2 it is evident that for models 2

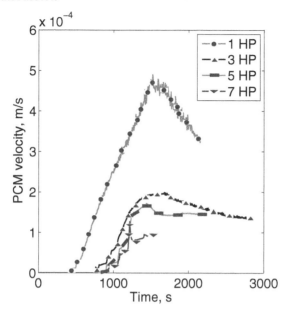

FIGURE 9.6
Transient variation of PCM velocity for the models considered [68].

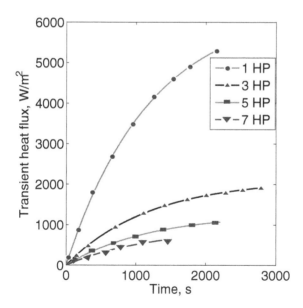

FIGURE 9.7
Transient heat flux variation of condenser wall for the models considered [68].

TABLE 9.2
Variation of temperature and melt fraction with increase in number of HPs
[68]

Model ID	Number of Hps	v_h	Temperature after 1500s (K)	Melt fraction after 1500s
1	1	0.26	345	0.38
2	3	0.79	323	0.50
3	**5**	**1.31**	**319**	**0.72**
4	7	1.83	339	0.89
5	9	2.36	360	1.00

and 3, after 1500s only 50% and 70% of the PCM have melted respectively.
Natural convection plays a significant role in transfer of heat to the unmelted
PCM in a uniform pattern. This effect is reflected in the velocity plot seen in
Figure 9.6, when the velocities are seen to be almost constant from t=1200 to
1600s for 3 and 5 heat pipe cases. The melt fraction contours for the 3 heat
pipe case is shown in Figure 9.8.

The relatively constant velocity (Figure 9.6) observed during the melting
in case 5 contributes to the consistent heating of the solid PCM and shows
up the best performance. The gradual erosion of the unmelted solid layer in
the 5 HP case helps maintain the condenser wall temperature below the safe
limits (usually 333 K).

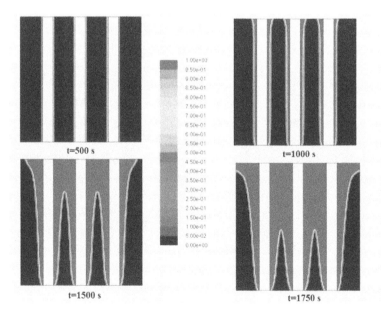

FIGURE 9.8
Contours of the volume fraction of the case with 3 HPs [68].

9.6 Conclusions

The melting process of a thermosyphon assisted LHTES system with n-eicosane as PCM enclosed by a square enclosure was numerically simulated using a synergistic transient numerical approach, wherein a lumped model was combined with full numerical simulations. The effects of heat pipe spacing, numbers and the effect of natural convection on the melting process were studied. From the numerical study, it was evident that natural convection plays a significant role in the melting process of PCM. Furthermore, it was seen that there exists an optimum ratio of heat pipe volume to PCM volume that enhances the performance of LHTES system. Therefore, increasing the number of HPs maintains the LHTES system under safe limit only up to a particular number, and for the problem under consideration this number is 5. The results obtained from this study and the previous chapters confirm that there is more than one option for the thermal designer to pick and choose in order to satisfy the design objectives.

9.7 Closure

This chapter examined the potential use of heat pipe as an alternative to fins in a latent heat thermal storage system. The next chapter presents the broad conclusions of the present and lists suggestions for future work.

10

CONCLUSIONS AND SCOPE FOR FUTURE WORK

10.1 Introduction

This book focused on the thermal performance of different types of PCM-based composite heat sinks through both experimental and numerical investigations. The experimental results gave an insight into the pertinent factors affecting the phase change process in the sensible and latent heating region such as power level, volume fraction of the PCM, volume fraction of the TCE, fin thickness, ambient temperature and so on. This information was helpful for mathematical modeling of the phase change phenomena with a view to determining the optimum configuration of the heat sink that maximizes the thermal performance along with salient conclusions on different heat sink configurations. The key departure of this study from most other works in the literature is the simultaneous consideration of heating and cooling cycles. The novelty of this work aside of this is its full fledged experiments on a rotating heat sink and characterization of its performance.

The book started with an introduction to the field of thermal management, phase change materials, thermal conductivity enhancers and thermal optimization techniques. The scarcity of the optimization strategies based on experimental data leads to the need for conducting experiments to characterise the thermal performance of phase change materials based heat sinks. The availablity of state of the art optimization algorithms, many of which are data driven, opens up vistas for the optimal design of heat sinks. **Chapter 2** provided a critical review of pertinent literature of the current research followed by a statement of the objectives of present study. Characterization of the PCMs and TCEs was briefly presented in **Chapter 3**. In **Chapter 4**, details of the experimental setup and instrumentation required to conduct experiments were presented. **Chapter 5** reported the results of the performance studies and thermal optimization of 72 pin fin heat sinks. Multi-objective optimization was performed and the results were reported in **Chapter 6**. The results of experimental investigations and the geometric optimization of matrix type heat sinks using a hybrid ANN−GA approach were discussed in **Chapter 7**. This chapter also provided a comparison of the thermal performance of the same TCE volume fraction of the no fin, matrix type fin and pin fin heat sinks. **Chapter 8** presented the results of the performance studies on cylindrical heat sinks subject to different fill ratio, orientation and rotation simultaneously. **Chapter 9** presented the results of the synergisitic numerical approach of heat sinks with thermosyphon as potential TCEs.

The major motivation behind all the studies carried out in this book can be seen from Figure 10.1. One can see that for any configuration heat sink, Figure 10.1a and Figure 10.1b, or even across multiple configurations, (Figure 10.1c) the melting and solidification are in high conflict. The '+' indicates the most desired ideal solution, which does not exist in real time. The '-' refers to most undesirable solution that occurs in real time. Any designer would want to maximize his/her solution from '-' and get closer to '+'. Notwithstanding this, pulling out the "true optimum" is like looking for a needle in a haystack is not trivial and requires guidance from robust optimization strategies. A robust and exhaustive multi-objective optimization is required to resolve the trade offs and no single solution can be arrived at that satisfies both the objectives.

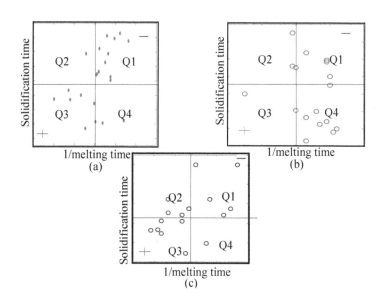

FIGURE 10.1
Conflict between melting and solidification time for the case of (a) 72 pin fin heat sink (b) matrix pin fin heat sink (c) cylindrical heat sink.

Based on the detailed investigations carried out in this work, the following broad conclusions are arrived at.

10.2 Major conclusions of the present study

1. From the initial experimental investigations it is evident that the PCM-based finned heat sinks show 10 times superior performance compared to the air-based finned/unfinned heat sinks. Furthermore, for an unfinned heat sink, a higher fill ratio does not ensure superior performance.

2. From the experimental investigation of a 72 pin fin heat sink subject to discrete heating, it is clear that the thermal distribution of the heat sources played a vital role in the performance of the heat sink. A case that showed superiority during the charging cycle, did not show the same superiority during the discharging cycle.

3. For the optimization problems of this class, a NSGA-II multi-objective algorithm is found to be superior over the other algorithms, and was hence was used to the problems that followed.

4. Experiments conducted on a new breed "matrix pin fin" type heat sink revealed that conducting elements in all three directions could induce quick melting of PCM leading to smaller charging time. However, the presence of conducting elements aided quicker solidification. Initially, the 72 pin fin was performing better compared to a matrix pin fin in terms of melting. Furthermore, geometric optimization of the heat sinks revealed that spacing between the fins played a major role in melting and the solidification heat transfer.

5. From the experimental investigation conducted on the cylindrical heat sinks, it was evident that for a lower volume of PCMs, the rotation and the orientation of the heat sink had positive effect on the melting heat transfer. However, for an air-based heat sink, the rotation of the heat sink helped in reducing the cooling time by 35%.

6. From the numerical study of thermosyphon assisted melting of PCM it was seen that there exists an optimum ratio of the heat pipe volume to the PCM that enhances the performance of the latent heat thermal energy storage system. Therefore, increasing the number of heat pipes helps maintains the latent heat thermal energy storage system under safe limits only up to a particular number, and for the problem under consideration this number is 5.

10.3   Suggestions for future work

Challenges are always associated with the thermal management of electronics in order to arrive at optimized designs. In order to address the growing thermal management issues, it is required that the present work be carried forward. Some of the possible directions are as follows :

1. Extension to the thermal management of devices with multiple phase change materials ([3] and [67]).

2. Performance studies on rotating heat sinks reported in Chapter 8 can be extended to higher rotational speeds and different phase change materials.

3. The experimental investigations can also be done for Gallium as a PCM.

4. Other than the organic PCMs such as n-eicosane, investigations can be done with metallic PCMs to effectively utilize the higher thermal conductivity than the organic PCMs. However, the weight penalty associated with the metallic PCMs has to be accounted for low weight heat sink designs.

5. Hybrid heat sinks that employ a heat pipe and a heat sink together may help us achieve the twin objectives of higher charging and lower discharging time.

10.4   Closure

This chapter gave a broad overview of this book and outlined the key conclusion of this study. Finally, the scope for future work was also presented.

Bibliography

[1] Ahmed, F., Deb, K., and Jindal, A. (2013). Multi-objective optimization and decision making approaches to cricket team selection. *Applied Soft Computing*, 13(1):402–414.

[2] Akhilesh, R., Narasimhan, A., and Balaji, C. (2005). Method to improve geometry for heat transfer enhancement in PCM composite heat sinks. *International Journal of Heat and Mass Transfer*, 48(13):2759–2770.

[3] Amritha, E., Srikanth, R., and Balaji, C. (2018). Experimental investigation of the thermal performance of matrix plate fin heat sink with multiple phase change materials. In *International Heat Transfer Conference Digital Library*. Begel House Inc.

[4] Asadi, E., da Silva, M. G., Antunes, C. H., Dias, L., and Glicksman, L. (2014). Multi-objective optimization for building retrofit: A model using genetic algorithm and artificial neural network and an application. *Energy and Buildings*, 81(0):444–456.

[5] Auger, A., Bader, J., Brockhoff, D., and Zitzler, E. (2009). Theory of the hypervolume indicator: optimal μ-distributions and the choice of the reference point. In *Proceedings of the Tenth ACM SIGEVO Workshop on Foundations of Genetic Algorithms*, pages 87–102. ACM.

[6] Baby, R. and Balaji, C. (2012). Experimental investigations on phase change material based finned heat sinks for electronic equipment cooling. *International Journal of Heat and Mass Transfer*, 55(5-6):1642–1649.

[7] Baby, R. and Balaji, C. (2013). Thermal optimization of pcm based pin fin heat sinks: an experimental study. *Applied Thermal Engineering*, 54(1):65–77.

[8] Baby, R. and Balaji, C. (2019). *Thermal Management of Electronics, Volume I: Phase Change Material-Based Composite Heat Sinks—An Experimental Approach*. Momentum Press.

[9] Bae, J. H. and Hyun, J. M. (2004). Time-dependent buoyant convection in an enclosure with discrete heat sources. *International Journal of Thermal Sciences*, 43(1):3–11.

[10] Bairi, A. and de Maria, J. G. (2013). Nu–ra–fo correlations for transient free convection in 2d convective diode cavities with discrete heat sources. *International Journal of Heat and Mass Transfer*, 57(2):623–628.

[11] Balaji, C. (2011). *Essentials of Thermal System Design and Optimization.* Ane Books Pvt.

[12] Bejan, A. (2013). *Convection Heat Transfer.* John Wiley.

[13] Boggs, P. T. and Tolle, J. W. (1995). Sequential quadratic programming. *Acta Numerica*, 4:1–51.

[14] Charnes, A. and Cooper, W. W. (1977). Goal programming and multiple objective optimizations: Part 1. *European Journal of Operational Research*, 1(1):39–54.

[15] Chaudhry, S. B., Hung, V. C., Guha, R. K., and Stanley, K. O. (2011). Pareto-based evolutionary computational approach for wireless sensor placement. *Engineering Applications of Artificial Intelligence*, 24(3):409–425.

[16] Cho, E. S., Koo, J.-M., Jiang, L., Prasher, R. S., Kim, M. S., Santiago, J. G., Kenny, T. W., and Goodson, K. E. (2003). Experimental study on two-phase heat transfer in microchannel heat sinks with hotspots. In *Semiconductor Thermal Measurement and Management Symposium, 2003. Ninteenth Annual IEEE*, pages 242–246. IEEE.

[17] Cuco, A. P. C., de Sousa, F. L., Vlassov, V. V., and da Silva Neto, A. J. (2011). Multi-objective design optimization of a new space radiator. *Optimization and Engineering*, 12(3):393–406.

[18] Das, B. and Giri, A. (2014). Second law analysis of an array of vertical plate-finned heat sink undergoing mixed convection. *International Communications in Heat and Mass Transfer*, 56(0):42–49.

[19] Deb, K., Pratap, A., Agarwal, S., and Meyarivan, T. (2002). A fast and elitist multiobjective genetic algorithm: Nsga-ii. *Evolutionary Computation, IEEE Transactions on*, 6(2):182–197.

[20] Demuth, H., Beale, M., and Hagan, M. (1992). Neural network toolbox. *For Use with MATLAB. The MathWorks Inc*, 2000.

[21] Dominguez, M., Fernandez-Cardador, A., Cucala, A. P., Gonsalves, T., and Fernández, A. (2014). Multi objective particle swarm optimization algorithm for the design of efficient ato speed profiles in metro lines. *Engineering Applications of Artificial Intelligence*, 29:43–53.

[22] Dubovsky, V., Barzilay, G., Granot, G., Ziskind, G., and Letan, R. (2009). Study of pcm-based pin-fin heat sinks. In *ASME 2009 Heat Transfer*

Summer Conference collocated with the InterPACK09 and 3rd Energy Sustainability Conferences, pages 857–863. American Society of Mechanical Engineers.

[23] El-Wahed, W. F. A. and Lee, S. M. (2006). Interactive fuzzy goal programming for multi-objective transportation problems. *Omega*, 34(2):158–166.

[24] Fan, L.-W., Xiao, Y.-Q., Zeng, Y., Fang, X., Wang, X., Xu, X., Yu, Z.-T., Hong, R.-H., Hu, Y.-C., and Cen, K.-F. (2013). Effects of melting temperature and the presence of internal fins on the performance of a phase change material (pcm)-based heat sink. *International Journal of Thermal Sciences*, 70(0):114–126.

[25] Faraji, M. and El Qarnia, H. (2010). Numerical study of free convection dominated melting in an isolated cavity heated by three protruding electronic components. *Components and Packaging Technologies, IEEE Transactions on*, 33(1):167–177.

[26] Farsi, H., Joly, J.-L., Miscevic, M., Platel, V., and Mazet, N. (2003). An experimental and theoretical investigation of the transient behavior of a two-phase closed thermosyphon. *Applied Thermal Engineering*, 23(15):1895–1912.

[27] Fok, S. C., Shen, W., and Tan, F. L. (2010). Cooling of portable handheld electronic devices using phase change materials in finned heat sinks. *International Journal of Thermal Sciences*, 49:109–117.

[28] Furtuna, R., Curteanu, S., and Leon, F. (2011). An elitist non-dominated sorting genetic algorithm enhanced with a neural network applied to the multi-objective optimization of a polysiloxane synthesis process. *Engineering Applications of Artificial Intelligence*, 24(5):772–785.

[29] Greenspan, J. D., Roy, E. A., Caldwell, P. A., and Farooq, N. S. (2003). Thermosensory intensity and affect throughout the perceptible range. *Somatosensory and Motor Research*, 20(1):19–26.

[30] Halelfadl, S., Adham, A. M., Mohd-Ghazali, N., Mara, T., Estella, P., and Ahmad, R. (2014). Optimization of thermal performances and pressure drop of rectangular microchannel heat sink using aqueous carbon nanotubes based nanofluid. *Applied Thermal Engineering*, 62(2):492–499.

[31] Ho, C.-J. and Viskanta, R. (1984). Heat transfer during melting from an isothermal vertical wall. *ASME,Journal of Heat Transfer*, 106(1):12–19.

[32] Hosseinizadeh, S., Tan, F., and Moosania, S. (2011). Experimental and numerical studies on performance of pcm-based heat sink with different configurations of internal fins. *Applied Thermal Engineering*, 31(17-18):3827–3838. 2010 Special Issue.

[33] Huang, C.-H., Lu, J.-J., and Ay, H. (2011). A three-dimensional heat sink module design problem with experimental verification. *International Journal of Heat and Mass Transfer*, 54(7-8):1482–1492.

[34] Huang, W. (2007). *Hotspot: A Chip and Package Compact Thermal Modeling Methodology for VLSI Design*, volume 67.

[35] Humphries, W. R. and Griggs, E. I. (1977). A design handbook for phase change thermal control and energy storage devices. Technical report, National Aeronautics and Space Administration, Huntsville, AL (USA). George C. Marshall Space Flight Center.

[36] Husain, A. and Kim, K.-Y. (2008). Optimization of a microchannel heat sink with temperature dependent fluid properties. *Applied Thermal Engineering*, 28(89):1101–1107.

[37] Jang, D., Yook, S.-J., and Lee, K.-S. (2014). Optimum design of a radial heat sink with a fin-height profile for high-power led lighting applications. *Applied Energy*, 116(0):260–268.

[38] Jaworski, M. (2012). Thermal performance of heat spreader for electronics cooling with incorporated phase change material. *Applied Thermal Engineering*, 35:212–219.

[39] Jones, B. J., Sun, D., Krishnan, S., and Garimella, S. V. (2006). Experimental and numerical study of melting in a cylinder. *International Journal of Heat and Mass Transfer*, 49(15-16):2724–2738.

[40] Kennedy, J. and Eberhart, R. (1995). Particle swarm optimization. In *Proceedings of the IEEE international conference on neural networks*, 4:1942–1948. CiteSeer.

[41] Koplow, J. P. (2010). A fundamentally new approach to air-cooled heat exchangers. *Sandia Report No. SANDIA2010-0258*.

[42] Kozak, Y., Abramzon, B., and Ziskind, G. (2013). Experimental and numerical investigation of a hybrid pcm-air heat sink. *Applied Thermal Engineering*, 59(1-2):142–152.

[43] Krishnan, S., Garimella, S. V., and Kang, S. S. (2005). A novel hybrid heat sink using phase change materials for transient thermal management of electronics. *Components and Packaging Technologies, IEEE Transactions on*, 28(2):281–289.

[44] Leoni, N. and Amon, C. (1997). Transient thermal design of wearable computers with embedded electronics using phase change materials. *ASME-PUBLICATIONS-HTD*, 343:49–56.

[45] Levin, P. P., Shitzer, A., and Hetsroni, G. (2013). Numerical optimization of a pcm-based heat sink with internal fins. *International Journal of Heat and Mass Transfer*, 61:638–645.

[46] Liu, Z., Wang, Z., and Ma, C. (2006). An experimental study on the heat transfer characteristics of a heat pipe heat exchanger with latent heat storage. Part ii: Simultaneous charging/discharging modes. *Energy Conversion and Management*, 47(7-8):967–991.

[47] Luo, Z., Cho, H., Luo, X., and Cho, K. (2008). System thermal analysis for mobile phone. *Applied Thermal Engineering.*, 28:1889–1895.

[48] Mahmoud, S., Tang, A., Toh, C., AL-Dadah, R., and Soo, S. L. (2013). Experimental investigation of inserts configurations and PCM type on the thermal performance of PCM based heat sinks. *Applied Energy*, 112(0):1349–1356.

[49] Martin, V., He, B., and Setterwall, F. (2010). Direct contact pcm water cold storage. *Applied Energy*, 87(8):2652–2659.

[50] Nadarajah, S. K. and Tatossian, C. (2010). Multi-objective aerodynamic shape optimization for unsteady viscous flows. *Optimization and Engineering*, 11(1):67–106.

[51] Nayak, K., Saha, S., Srinivasan, K., and Dutta, P. (2006a). A numerical model for heat sinks with phase change materials and thermal conductivity enhancers. *International Journal of Heat and Mass Transfer*, 49(11-12):1833–1844.

[52] Nayak, K., Saha, S., Srinivasan, K., and Dutta, P. (2006b). A numerical model for heat sinks with phase change materials and thermal conductivity enhancers. *International Journal of Heat and Mass Transfer*, 49(11-12):1833–1844.

[53] Pakrouh, R., Hosseini, M., Ranjbar, A., and Bahrampoury, R. (2015). A numerical method for pcm-based pin fin heat sinks optimization. *Energy Conversion and Management*, 103:542–552.

[54] Pillai, K. and Brinkworth, B. (1976). The storage of low grade thermal energy using phase change materials. *Applied Energy*, 2(3):205–216.

[55] Riquelme, N., Von Lücken, C., and Baran, B. (2015). Performance metrics in multi-objective optimization. In *Computing Conference (CLEI), 2015 Latin American*, pages 1–11. IEEE.

[56] Robak, C. W., Bergman, T. L., and Faghri, A. (2011). Enhancement of latent heat energy storage using embedded heat pipes. *International Journal of Heat and Mass Transfer*, 54(15-16):3476–3484.

[57] Sanaye, S. and Hajabdollahi, H. (2010). Thermal-economic multi-objective optimization of plate fin heat exchanger using genetic algorithm. *Applied Energy*, 87(6):1893–1902.

[58] Schittkowski, K. (1983). On the convergence of a sequential quadratic programming method with an augmented lagrangian line search function 2. *Optimization*, 14(2):197–216.

[59] Sertkaya, A. A., Bilir, A., and Kargici, S. (2011). Experimental investigation of the effects of orientation angle on heat transfer performance of pin-finned surfaces in natural convection. *Energy*, 36(3):1513–1517.

[60] Shabgard, H., Bergman, T., Sharifi, N., and Faghri, A. (2010). High temperature latent heat thermal energy storage using heat pipes. *International Journal of Heat and Mass Transfer*, 53(15-16):2979–2988.

[61] Shatikian, V., Ziskind, G., and Letan, R. (2008). Heat accumulation in a pcm-based heat sink with internal fins. In *Eurotherm-5th European Thermal-Sciences Conference, Netherlands*.

[62] Sridharan, S., Srikanth, R., and Balaji, C. (2018). Multi objective geometric optimization of phase change material based cylindrical heat sinks with internal stem and radial fins. *Thermal Science and Engineering Progress*, 5:238–251.

[63] Srikanth, R. and Balaji, C. (2014a). Experimental investigation of thermal performance of phase change material based composite heat sinks with discrete heat sources. In *International Heat Transfer Conference Digital Library*. Begel House Inc.

[64] Srikanth, R. and Balaji, C. (2014b). Experimental investigation of thermal performance of phase change material based composite heat sinks with discrete heat sources. *Proceedings of the 15th International Heat Transfer Conference, IHTC-15, 2014, Kyoto*.

[65] Srikanth, R. and Balaji, C. (2017a). Experimental investigation on the heat transfer performance of a PCM based pin fin heat sink with discrete heating. *International Journal of Thermal Sciences*, 111:188–203.

[66] Srikanth, R. and Balaji, C. (2017b). Heat transfer correlations for a composite pcm based 72 pin fin heat sink with discrete heating at the base. *INAE Letters*, 2(3):65–71.

[67] Srikanth, R., Balaji, C., et al. (2018). Experimental investigation on the thermal performance of a pcm based cylindrical heat sink with multiple pcms. In *International Heat Transfer Conference Digital Library*. Begel House Inc.

[68] Srikanth, R., Nair, R. S., and Balaji, C. (2016). Thermosyphon assisted melting of pcm inside a rectangular enclosure: A synergistic numerical approach. In *Journal of Physics: Conference Series*, volume 745, page 032130. IOP Publishing.

[69] Srikanth, R., Nemani, P., and Balaji, C. (2015a). Multi-objective geometric optimization of a pcm based matrix type composite heat sink. *Applied Energy*, 156:703–714.

[70] Srikanth, R., Nemani, P., and Balaji, C. (2015b). Multi-objective geometric optimization of a pcm based matrix type composite heat sink. *Applied Energy*, 156:703–714.

[71] Stupar, A., Drofenik, U., and Kolar, J. W. (2012). Optimization of phase change material heat sinks for low duty cycle high peak load power supplies. *Components, Packaging and Manufacturing Technology, IEEE Transactions on*, 2(1):102–115.

[72] Sun, X., Zhang, Q., Medina, M. A., Liu, Y., and Liao, S. (2014). A study on the use of phase change materials (pcms) in combination with a natural cold source for space cooling in telecommunications base stations (tbss) in China. *Applied Energy*, 117(0):95–103.

[73] Tari, I. and Mehrtash, M. (2013). Natural convection heat transfer from inclined plate-fin heat sinks. *International Journal of Heat and Mass Transfer*, 56(1-2):574–593.

[74] Tiari, S., Qiu, S., and Mahdavi, M. (2015). Numerical study of finned heat pipe-assisted thermal energy storage system with high temperature phase change material. *Energy Conversion and Management*, 89:833–842.

[75] Voller, V. and Prakash, C. (1987a). A fixed grid numerical modelling methodology for convection-diffusion mushy region phase-change problems. *International Journal of Heat and Mass Transfer*, 30(8):1709–1719.

[76] Voller, V. R. and Prakash, C. (1987b). A fixed grid numerical modelling methodology for convection-diffusion mushy region phase-change problems. *International Journal of Heat and Mass Transfer*, 30(8):1709–1719.

[77] Wang, X.-Q., Yap, C., and Mujumdar, A. S. (2008). A parametric study of phase change material (pcm)-based heat sinks. *International Journal of Thermal Sciences*, 47(8):1055–1068.

[78] Wang, Y., Amiri, A., and Vafai, K. (1999). An experimental investigation of the melting process in a rectangular enclosure. *International Journal of Heat and Mass Transfer*, 42(19):3659–3672.

[79] Wang, Y.-H. and Yang, Y.-T. (2011). Three-dimensional transient cooling simulations of a portable electronic device using PCM (phase change materials) in multi-fin heat sink. *Energy*, 36(8):5214–5224. PRES 2010.

[80] Webb, B. and Viskanta, R. (1985). On the characteristic length scale for correlating melting heat transfer data. *International Communications in Heat and Mass Transfer*, 12(6):637–646.

[81] Weng, Y.-C., Cho, H.-P., Chang, C.-C., and Chen, S.-L. (2011a). Heat pipe with {PCM} for electronic cooling. *Applied Energy*, 88(5):1825–1833.

[82] Yazici, M. Y., Avci, M., and Aydin, O. (2019). Combined effects of inclination angle and fin number on thermal performance of a pcm-based heat sink. *Applied Thermal Engineering*, page 113956.

[83] Yin, H., Gao, X., Ding, J., Zhang, Z., and Fang, Y. (2010). Thermal management of electronic components with thermal adaptation composite material. *Applied Energy*, 87(12):3784–3791.

[84] Yoon, K. P. and Hwang, C.-L. (1995). *Multiple Attribute Decision Making: an Introduction*, volume 104. Sage Publications.

[85] Zaman, F. S., Turja, T. S., and Molla, M. M. (2013). Buoyancy driven natural convection flow in an enclosure with two discrete heating from below. *Procedia Engineering*, 56:104–111.

[86] Zhang, Y., Chen, Z., Wang, Q., and Wu, Q. (1993). Melting in an enclosure with discrete heating at a constant rate. *Experimental Thermal and Fluid Science*, 6(2):196–201.

[87] Zheng, N. and Wirtz, R. (2004). A hybrid thermal energy storage device, part 1: design methodology. *ASME, Journal of Electronic Packaging*, 126(1):1–7.

[88] Zhou, D., Zhao, C., and Tian, Y. (2012). Review on thermal energy storage with phase change materials (pcms) in building applications. *Applied Energy*, 92(0):593–605.

Index